...version and S...

A Guide to Principles

and Practice

A beautiful example of tree culture

Scots pine in a Forestry Commission plantation

The Conversion and Seasoning of Wood

William H. Brown

F.I.W.Sc., A.M.I.W.M.

STOBART DAVIES LTD
HERTFORD

British Library Cataloguing in Publication Data
Brown, William Henry
The conversion & seasoning of wood : a guide to principles and practice.
1. Timber
I. Title
674 TA419

ISBN 0–85442–037–1

Published 1988, reprinted 1995 by Stobart Davies Ltd., Priory House, Priory Street, Hertford SG14 1RN

Printed and bound in Great Britain by BPC Wheatons Ltd., Exeter

Acknowledgements

In preparing this book I have been greatly helped by the wide circle of knowledgeable people it has been my good fortune to meet, or have contact with, both at home and abroad, in a lifetime spent in the timber and woodworking trades. Friendly criticism, constructive advice, and encouragement have come my way and I acknowledge my sincere gratitude to all of them. I am indebted to Hugh Blogg, a craftsman whose work I have long admired, for supplying photographs of some of his burr veneer work, and I want to thank John Arrowsmith for filling in the gaps in my knowledge of the Mini-Seasoner he developed for Ebac. I am especially grateful to Leslie Cubbage for the time spent in supplying information and for obtaining necessary photographs. Grateful acknowledgement is made of the help and advice on the subject of on-site sawing given by Mr E.N. Visram, Managing Director of Forest and Sawmill Equipments. I am grateful too, for the data and photographs supplied by Peter Chen of Southern Illinois University. I want also to thank Ernest G. Gobert of Protimeter, and Nicholas Austin of Brannan Thermometers for their advice and for supplying photographs. Finally, special thanks are extended to Associate Professor Eugene Wengert of Virginia Polytechnic Institute and State University for his support now, and in the past, on the matter of seasoning technology. I am extremely grateful for his permission to quote from his research work including construction details of his solar dryer for small business. For all the kind assistance I have been given, this work has been a pleasurable pursuit.

Publisher's Note

Due to the Author's untimely death, his manuscript was not in its final form when delivered to the Publisher. We are most grateful to the well respected timber consultant, Ron Hooks, for editing the proofs and adding new material, where required, to ensure that this book offers to all concerned the most relevant information on current technology and practice in the conversion and seasoning of wood.

Contents

Introduction

Efficient and careful conversion of timber from the log and its proper seasoning are at the heart of the successful wood product. No-one likes to receive complaints about their work, particularly when it is not so much the workmanship but the materials which cause the breakdown of an otherwise well-worked article.

In the following chapters the many aspects of effective material preparation will be studied in detail, from the reduction of waste caused by poor conversion and machining and inadequate and/or incorrect seasoning to the various methods of bringing timber to the right moisture content level for the situation in hand.

One of the most important economic factors for the wood user is how much good, usable timber he can get from any log and having extracted that timber, how much or how little waste there will be in the seasoning process.

The waste factor looms large for the unwary, ranging from the development of splitting and warping to insect damage and staining. These and many other defects are considered and remedies offered.

In many instances, the consumer uses timber dried in a kiln or dehumidifier, but it does not automatically follow that this provides a complete answer to seasoning, although it goes a long way towards doing so. The reason is basically that a kiln or dehumidifier will only produce what is asked of it and in a good many cases the customer does not always appreciate the extent of the variation in kiln dried timber and the required condition to suit a particular situation. Much effort goes into the kilning of timber but if the specified final moisture content is at a level out of balance with what is really needed then the efficiency of the service could be in doubt.

Logically, any study of wood must commence with an appraisal of its structure, but for most practical purposes a highly specialised knowledge is not essential. However, since the nature and properties of wood constitute the basis of all wood use, it is essential that rudimentary knowledge be increased if the most satisfactory results are to be obtained from a material designed by nature and not engineered by man.

Nature did not design a tree with altruistic motives in mind; each different element and tissue contained in a woody stem was manufactured to meet a special activity or to perform a particular function concerned with growth. When man decides to fell a tree for his own use he must, in consequence, learn to recognise those desirable properties nature has provided and adapt them to advantage; but he must also learn how to modify other properties which, so essential to plant growth, are less essential now.

This is particularly relevant to the small user of wood who frequently acquires odd parcels of wood, some well-known, others of obscure or doubtful origin, sometimes in low-grade condition. He may be offered an isolated tree, standing or freshly felled, at any time of year, of a species which might range from a fruit or nut tree to a hybrid decorative tree which has outlived its usefulness and appeal.

Trees are sometimes attacked by beetle pests and fungal organisms which may be reasons for disposal by the owner. The woodworker may be offered wood containing decay which, depending on an individual's point of view may be regarded as decorative. In industry, decay in wood is invariably a cause for rejection, while on the other hand, beauty is in the eye of the beholder and it is often prized by the craftsman.

Decay in wood is variously described as doaty or spalted, and is the result of a tree being attacked by one of many white rot fungi. Since these particular organisms attack and destroy both lignin and cellulose, much depends upon the degree of biodegradation in the wood at the time of use as to whether it will stand up to machining without undue tearing out of the attacked areas.

Wood, being a naturally occurring material, varies in its character according to species, and from tree to tree, and one of the principal elements in the life processes of a tree is water. So too, the greatest single element which leads to deterioration of wood as a material is also water; and deterioration by shrinking, swelling, splitting and distortion can be attributed to direct or consequential effect of moisture caused by (a) atmospheric and climatic action, (b) wetting and drying effects, and (c) accidental wetting, say by flooding.

Many anomalies appear to exist when wood is seen to deteriorate, and depending on the conditions prevailing at the time and the type of deterioration evident, it is often assumed the air is too wet, or the temperature too high, neither of which is strictly correct. Air may be too wet or too dry to suit wood in a given set of circumstances but these are measures of the surrounding air's capacity to hold moisture which in turn is affected by temperature changes. If timber is stacked in the open air in winter, no-one would want to use the wood immediately for furniture because the outdoor conditions would be considered wetter than those prevailing indoors. What is not usually realised is that cold winter air, if drawn into a living room and heated up, could now be exceedingly dry and detrimental to interior wood, that is, unless some moisture is added to the air.

In a practical sense, the outdoor wood is wet because of the average winter air conditions, but cold air can only hold a certain amount of water vapour which is why the wood dries little, if at all, in winter; but if that air is heated, either artificially, as in a warm room, or naturally, by the sun, it will then have the capacity to hold a lot more moisture. It is therefore important for wood users to understand as much as possible of the effect different atmospheres have on wood both indoors and outdoors.

The basic principles governing wood use are the same for the craftsmen as they are for the large manufacturer but in practice their roles differ somewhat. The professional users and specifiers of wood such as architects, building consortiums, furniture and joinery plants, etc., invariably demand their material in such a way as to ensure it fits nicely into standard techniques and procedures and in no way contravenes established rules, regulations and specifications. This requirement is met by suppliers who, from long experience, offer a service geared to these industrial practices and preferences; in short, both customer and supplier operate within a long-established and sophisticated set of rules.

On the other side of this well-ordered world is another, no less important group of wood user: the craftsman turner, carver, sculptor, toy maker, cabinet maker, shop fitter, small builder, student, and enthusiastic home woodworker devoting his or her spare time to a hobby based simply on a love of wood.

It is mainly for this second group that this book has been prepared, with a distinct bias toward the individual craftsmen and amateur woodworker, but with the hope that the pages that follow will serve as an aide mémoire to the professional, and as a work of reference to all wood users.

Botanical and anatomical references have been kept to a minimum so as not to confuse the general principles of the text.

1
Timber Conversion

Before a log is converted into timber some thought should be given to the method of sawing and to the type and species of wood present.

Wood is not a solid, homogeneous substance, but a heterogeneous conglomeration of large numbers of very small elements or cells, most of these vertically opposed in the tree and designed to carry out certain functions. Some of these cells have very thick walls and act as supporting tissue thereby providing strength and resilience while others with thinner walls contribute to liquid food conduction. A smaller proportion of cells are laid down horizontally and form wood rays whose duty is to convey and store food.

Basically, wood substance is a porous mass of tube-like material held together by amorphous molecules of lignin and, depending upon species and types, containing various chemicals, minerals, gums, resins, tannin, silica, and so on.

The growth processes of a tree involves a stem or trunk consisting of a pith, formed by the original shoot which has special growing cells at its tip thus enabling height to be attained during each growing season, and cells forming a complete sheath round its periphery; accordingly, the tree gains height and girth progressively, annual increments of wood being laid down in a cone-like form.

Side shoots which ultimately become branches and finally knots, initiate in the pith, but since these gradually become removed or shortened by various means at the lower level of the tree bole, it is possible to visualise a standing tree as representing several logs of reducing quality: i.e., butt, second length, perhaps a third length, and lops and tops.

Butts produce the cleanest wood, generally speaking, and since the base of trees tend to broaden it is often the practice to fell them from above soil level in order to produce relatively straight logs. This is logical with trees which have large buttresses or flutes, but with some trees such as single decorative specimens, and fruit and nut trees, it often pays to dig away the soil from around the base and then saw

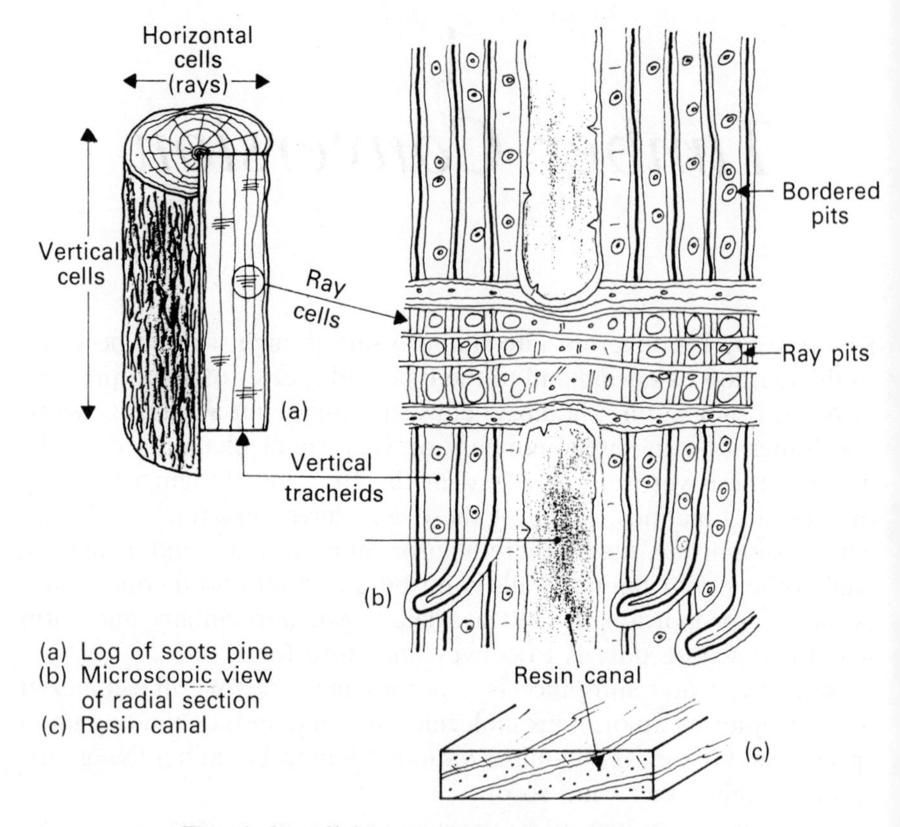

Fig. 1 It will be noted that the vertical tracheids conduct moisture through the length of the wood while the ray cells are horizontally opposed. Since the tracheid walls consist of linked molecules ray elements restrict movement radially. The pits permit moisture to move horizontally.

through at a point where the roots meet the bole. Walnut trees are usually grubbed out this way because the base of the bole frequently contains beautifully mottled or rippled wood.

Rootwood has no economic importance except in the case of the tree heath (*Erica arborea*) which grows in the Mediterranian area and is used for briar pipes. Rootwood of hardwood species is lighter in weight than the wood of the bole, but rootwood of conifers is usually similar in weight to the bole. Apart from objets d'art that are sometimes produced from the roots of trees, general use, say for

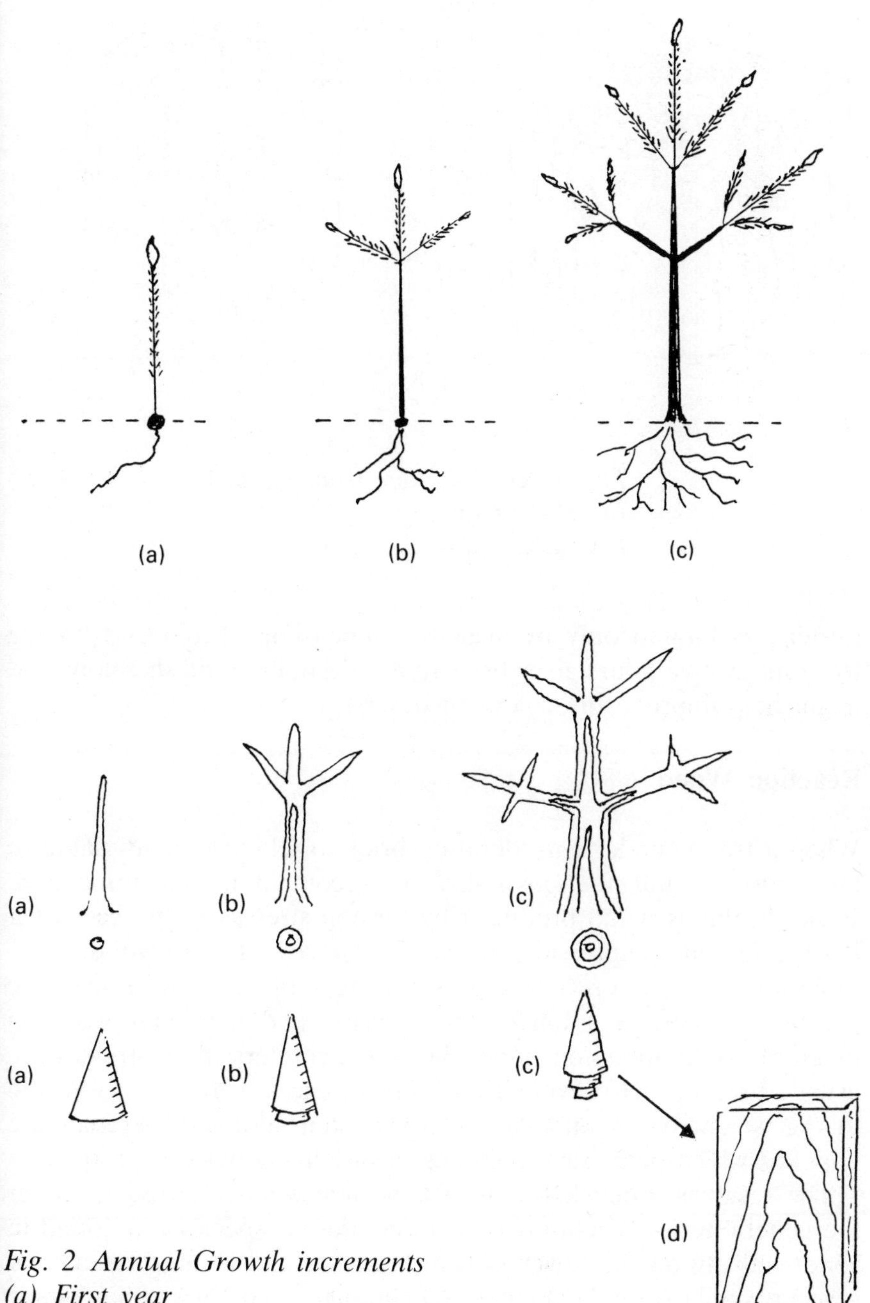

Fig. 2 Annual Growth increments
(a) First year
(b) Second year
(c) Third year
Wood laid down in cone-like formation (d)

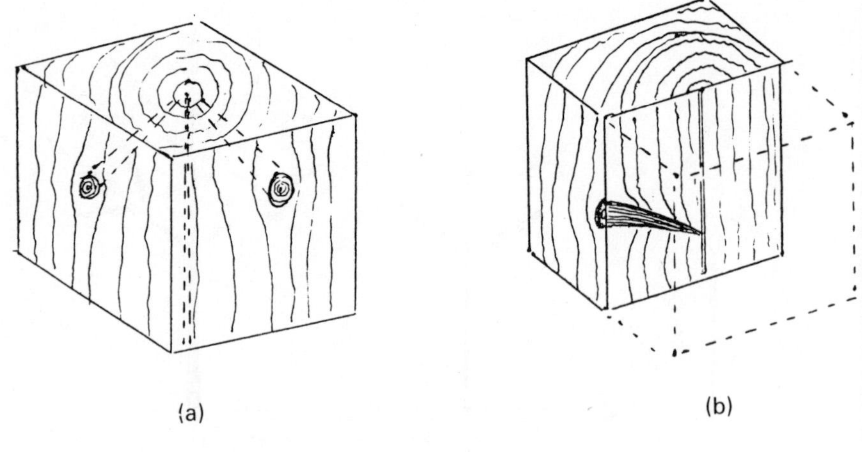

Fig. 3 Knots initiate from the pith
(a) Round knots
(b) Spike knot

turnery, is limited only by available dimensions. Prolonged boiling for some twelve hours gives briar root a desirable reddish colour and might help improve the colour of other species.

Reaction Wood

When a tree is under consideration prior to felling it is advisable to take into account the possibility of it containing reaction wood. Basically this is wood produced by tension stresses in the case of a hardwood and compression stresses in the case of a softwood.

With a perfectly erect tree its parts are subject to stress with the peripheral wood in a longitudinal tension and the inner wood in longitudinal compression; in smaller erect conifers these stresses are usually reversed. However, if a tree leans excessively or is subjected to the weight of an unbalanced crown then additional stresses are imposed which place severe strains on the living wood.

The result is a development of abnormal wood referred to under the generic term, reaction wood. In coniferous species it is found in zones, mostly on the lower side of branches and leaning boles, and it occurs similarly on the upper side in branches and boles of hardwood species.

Many trees offered to the craftsmen are gnarled or defective specimens and although capable of yielding a lot of good wood they require close inspection before deciding their potential value. For

Fig. 4 Knot formation pattern in Scots Pine. Each whorl represents one season's growth. Photograph, courtesy of TRADA

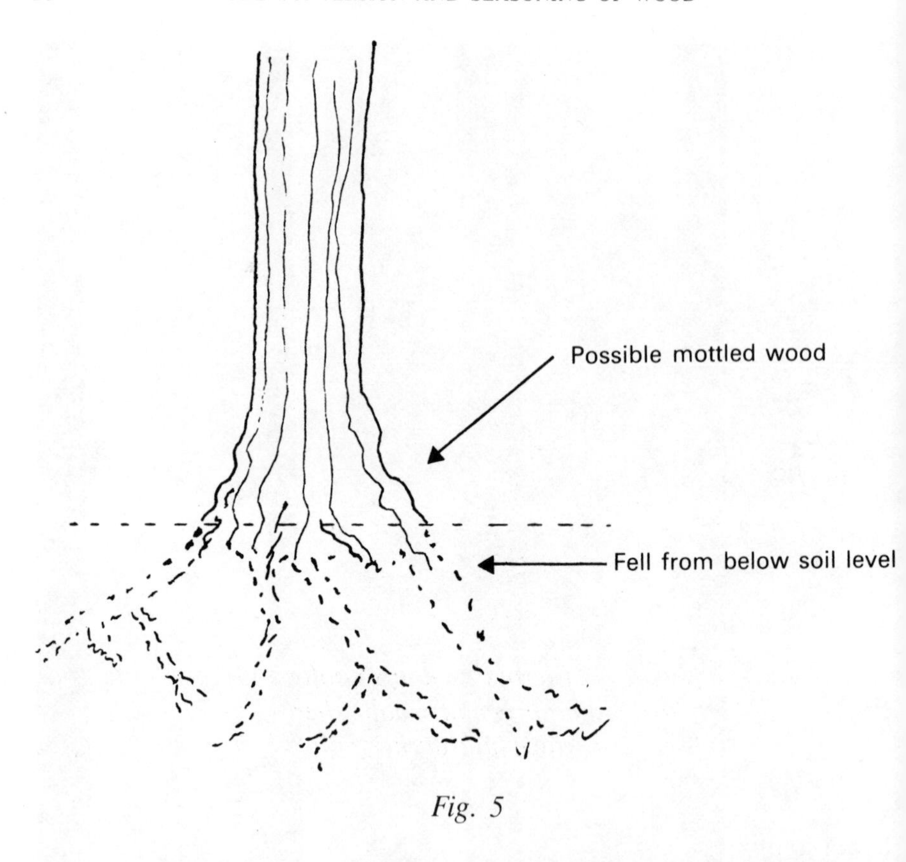

Fig. 5

instance, while the base of a tree might produce some nicely figured wood by virtue of the growth at this point it does not automatically follow that similar wood will be found if taken from below a heavy branch or from the branch itself, because this might not behave normally.

Compression wood in softwood species can readily be detected with the naked eye if the sawn-off end of a suspect branch is examined. Abnormal wood will be rather darker in colour than normal wood and will occur in zones of a reddish shade. This should not be confused with zones of incipient decay because these often appear reddish or reddish-brown in colour; however, they can be separated because discoloured zones attributable to decay often start from the periphery of the wood and broaden out as they proceed, whereas in compression wood there is a definite tendency for the darker colour to follow a radial path across the growth rings.

The pith is placed eccentrically, nearer the upper side of the branch, the growth rings are wider apart in the compression zone, and the

Fig. 6 Stem cross section of softwood showing compression wood where rings are wider and darker

wood is hard and brittle.

Tension wood in broad leaved trees shows an eccentric pith nearer the lower side with tension wood above. The wood may not appear much darker in colour than normal, but frequently it has a slightly lustrous, silvery appearance.

Reaction wood from any tree is brash, tending to split obliquely across the grain, while hardwood longitudinal surfaces work up woolly; it does not take kindly to any machine or finishing operation and it does have abnormally high longitudinal shrinkage, some 10 or 20 times greater than normal wood whose shrinkage along the grain is negligible. Since reaction wood occurs in zones, shrinkage can pull the wood very much out of shape.

Burrs

A burr or burl may be defined as a large woody excrescence on trunk, branch or root stock. The grain is highly contorted and gives rise to a characteristic type of figure. Burrs occur as a result of injury to the cambium which lies immediately under the bark; such injury may be caused by animals like rabbits, goats and deer, or by tractor or cart wheels, or by fire, but irritation of the cambium from which woody growth springs tends to cause groups of buds to develop in this area; the buds turn to so-called adventitious shoots while the surrounding woody tissue grows more rapidly than the tree wood as a whole. Annually, by an action known as cladoptosis, many of the thin shoots are shed, to leave a tiny pin knot, black or dark brown in colour in its place, but fresh shoots continue to appear year after year, with the shedding of twigs continuing.

As the excrescence develops the presence of the innumerable pin knots causes the fibres in the surrounding wood to twist considerably; the combination of knots and twisted grain being responsible for burr figure. Obviously, where the various elements in wood deviate severely from the normal, difficulties are presented, especially in seasoning, because normal shrinkage behaviour will not occur; typical burr growth will also affect the quality of sawn and finished wood.

Fig. 7 Burr oak slab ready for sawing to veneer 1·5mm thick in small sizes. Ivory rule shows 150mm. Photograph, courtesy of Hugh Blogg

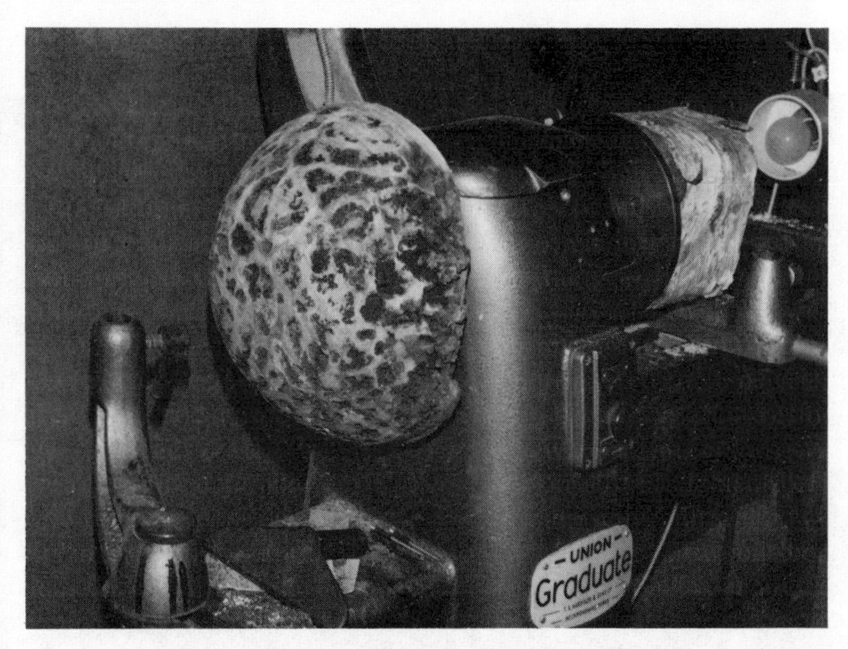

Figs. 8 & 9 A bowl being turned from oak burr. Photograph, courtesy of Hugh Blogg

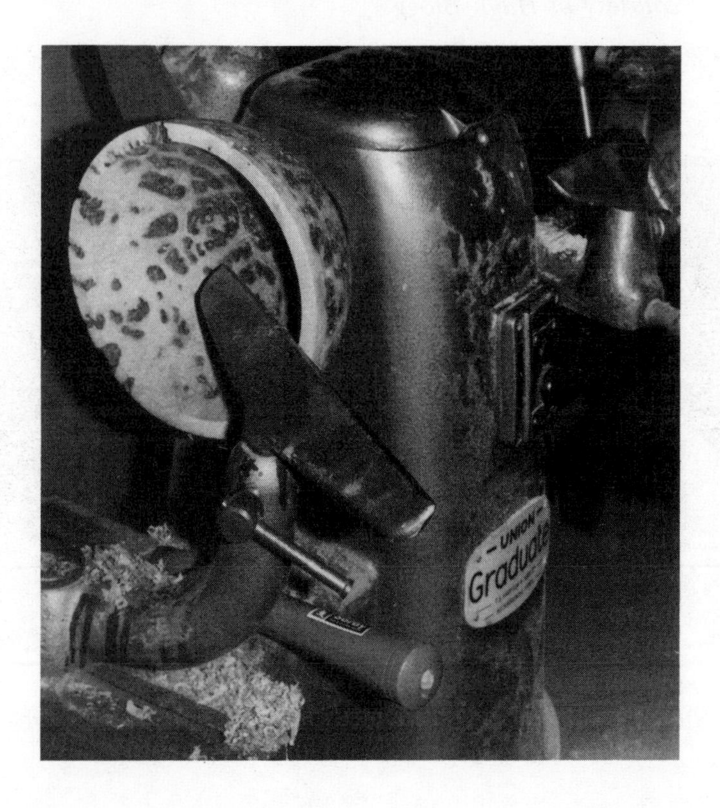

Fig. 10 Finished bowl in oak burr. Photograph, courtesy of Hugh Blogg

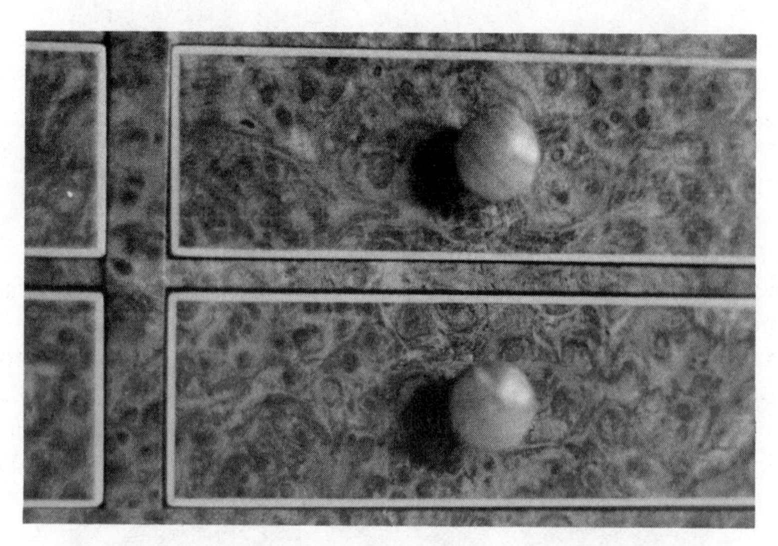

Fig. 11 Use of Thuya burr in miniature furniture making. Photograph, courtesy of Hugh Blogg

If a burr is sectioned tangentially, its figure will be made up of tiny pin knots in small clusters, but if sawn radially, it will take the form of spike knots; either way, the wood is highly stressed and great care is needed in attempting to produce sections which will remain flat, will not develop too many fine checks, and will accept suitable finishing media. Burrs may develop on many types of tree and especially on elm, oak, sycamore and maple.

Conversion

When a round log is converted into planks, boards and squares, in effect this is tantamount to squaring the circle; in other words, an organic mass in which all woody growth has taken place in circular fashion round a pith is now to be converted to rectangles or squares. Whatever form the conversion takes means the broad faces of a board, for example, will be at a particular angle with the axis of the log and will display the most prominent features of the wood structure at that angle and this might be to advantage or disadvantage to a proposed use.

A transverse section refers to an end grain surface, while a tangential section is defined as a lengthwise or longitudinal section in a plane tangential to a growth ring, i.e. at right angles to the radius: a radial section is a lengthwise or longitudinal section in a plane that passes through the pith, i.e. along the radius of a stem.

Depending upon the way in which a log is converted certain features and characteristics of the wood in question will be displayed in a particular way: if, for example, a log of oak is sawn radially its broad rays will present a very prominent silver grain figure on the wide surface of the board, the reason being that rays meander as long plates of tissue vertically in the tree and by sawing in a true radial plane patches of ray tissue become exposed.

A very prominent silver grain figure in oak is often deemed to be unacceptable in finished wood ware and frequently it can cause problems in machining because of the tendency for the true grain, vertically opposed to the horizontal ray elements, to tear out. When this occurs the wood is generally referred to as shelly.

If this same log is sawn a little off true radial, then the rays are angled away from the broad faces of the wood and a finer, raindrop figure is produced; if, on the other hand, the log was sawn tangentially,

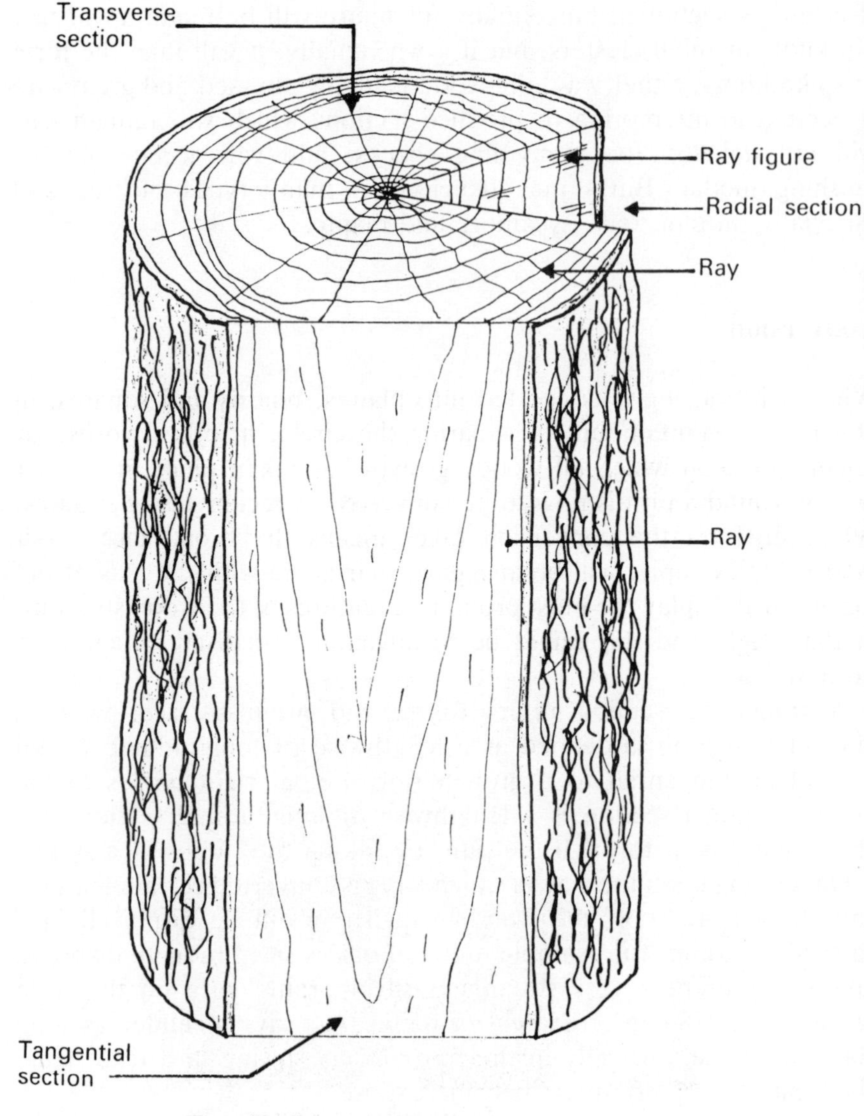

Fig. 12 Log sections:
Transverse = End Grain
Tangential = "Plain Sawn"
Radial = "Quarter Sawn" or "Rift Sawn".

the wide sawn faces of the boards would display the growth rings in a pattern corresponding to the cone-like growth structure, in other words, as part of an exaggerated series of inverted V's; the rays in this case would contribute nothing to appearance because they would

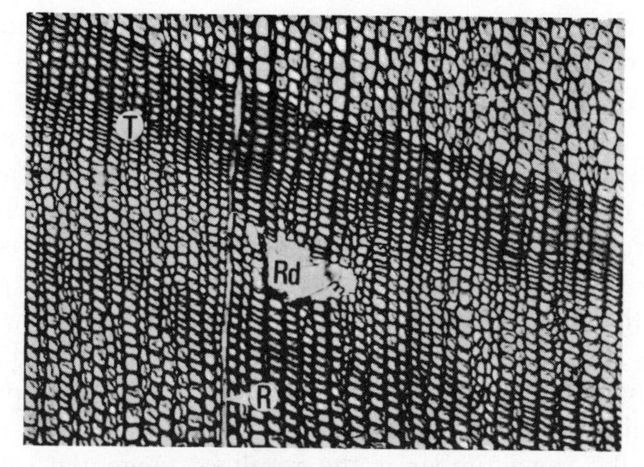

Fig. 13 Transverse section of a pine showing relatively simple structure magnification × 100
T = Tracheids
R = Rays
RD = Resin duct

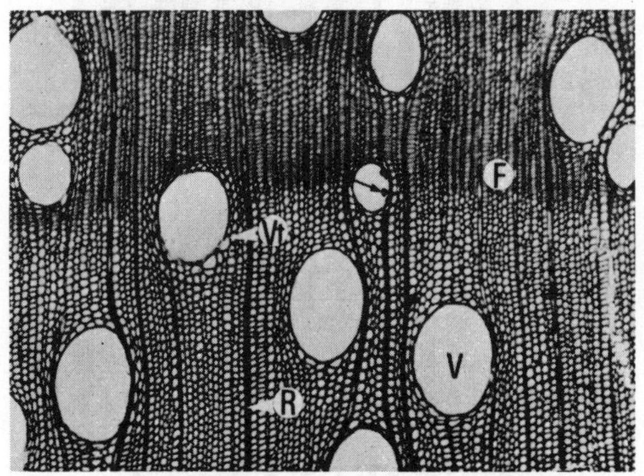

Fig. 14 Transverse section of a hardwood (Eucalyptus spp.) showing more complex structure × 10
V = Vessels
VT = Vasicentric tracheids
R = Rays
F = Fibres
Note, tyloses developing (arrow)

now represent a very large amount of narrow, boat shaped elements since the ends of each ray would be exposed.

Not all woods have prominent rays; the softwoods are examples, as are alder, spindlewood, tupelo gum, apple and pear among the hardwoods; and not all display prominent growth rings as do oak, elm, sweet chestnut and others, but even the most featureless woods show different characteristics according to how they are sawn.

Softwoods which display a darker-coloured latewood portion in each growth ring will produce a rather flowery figure when sawn tangentially, pitch pine and Douglas fir are examples; but all softwoods with clearly defined growth rings will look relatively plain, with straight grain when sawn radially.

Plain or flat sawn timber is the result of tangential sawing, and quarter sawn, rift sawn or edge grain timber is produced by sawing

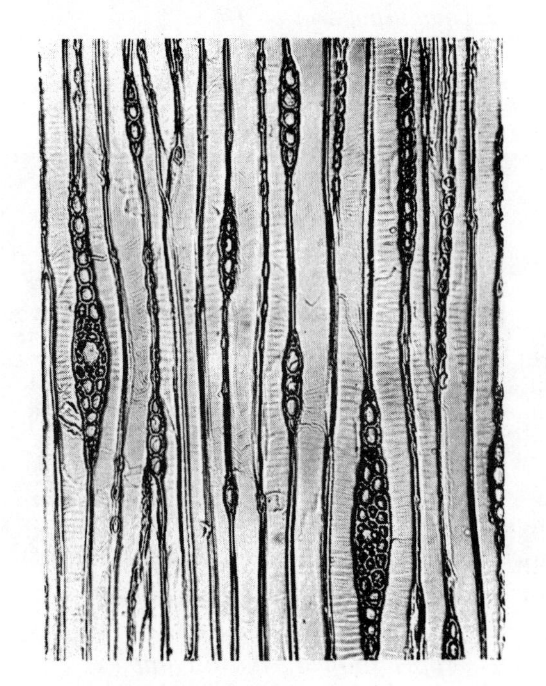

Fig. 15 Tangential longitudinal section of Douglas fir showing rays, two of which have central resin canals (magnification × 140)

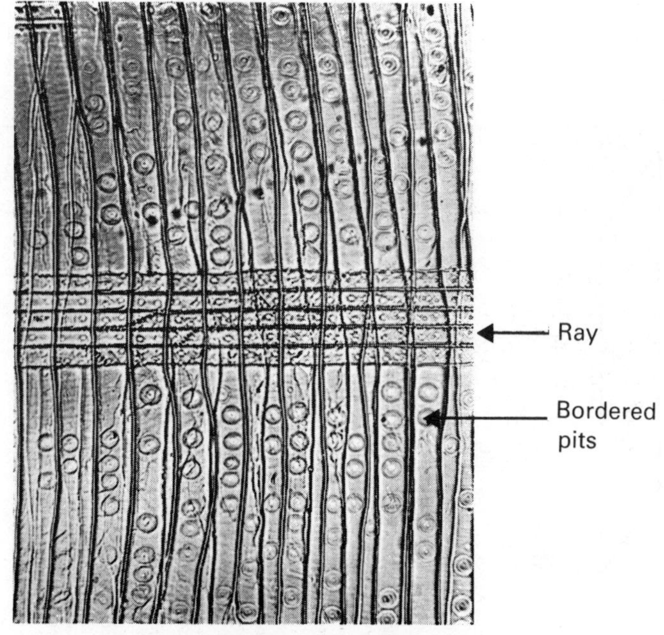

Fig. 16 Radial section of Douglas fir (× 140)

radially; however both terms need clarification if they are to be taken literally.

The most common and straightforward method of conversion is that termed through and through, whereby the log is sawn vertically or horizontally, depending on the type of saw used, by a succession of cuts. This is, in theory at least, tangential sawing, and largely the boards sawn from the centre of the log; these will be true quarter sawn because the plane of the cut passes through the pith, whereas the remainder of the boards will have their sawn faces in a plane tangential to the growth rings. In practice, some selecting out of the fully quartered stock would take place during grading, but in a non-commercial sense it is important to recognise that sawing out a log one way will produce different looking wood than that produced by some other form of sawing.

Technically, it is possible to convert a hardwood log so as to produce all true quarter sawn boards, i.e. radially sawn, but economically it is usually not a feasible proposition because of the time involved in handling the part log or quarter. Fig. 18b and c illustrates the problem. Quarter sawing dates back to a period when

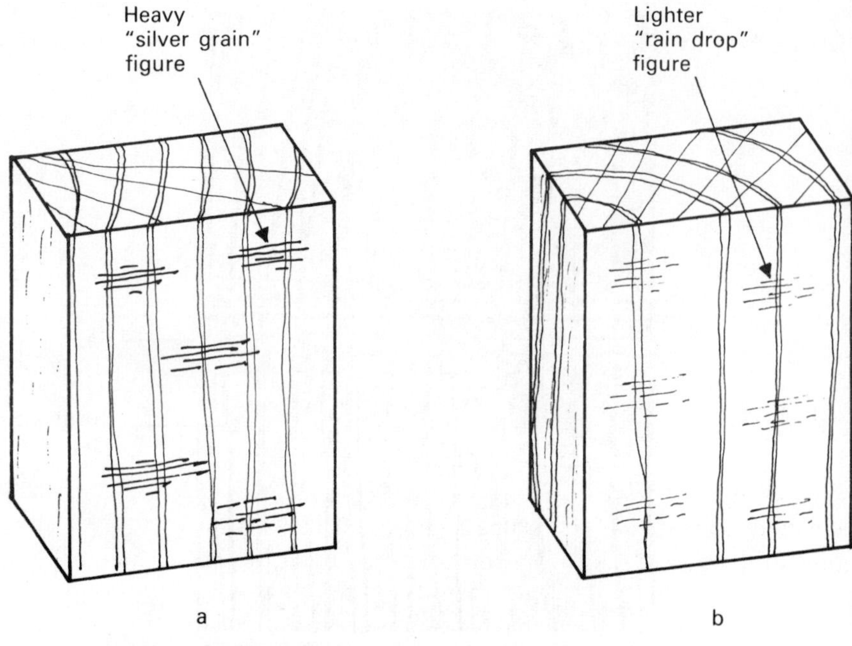

Fig. 17 Ray figure in oak
(a) True quarter-sawn wood gives prominent figure; sometimes difficult to finish cleanly
(b) Semi-quarter-sawn; figure is more subdued and less likely to tear in finishing: this applies to all species with large rays.

labour was relatively cheap in the thirteenth century. The Baltic countries were shipping to the UK and elsewhere, enormous quantities of oak to meet the demand for panelling and wainscots, and indeed for coffins, the boards for these purposes having to meet a width specification of at least 11″ (280mm) free from heart which was boxed out during conversion; large logs were therefore required. Even so, the amount of true quarter sawn, or more specifically, radial sawn wainscots that were produced was relatively small per log, Fig. 19.

In modern terms, hardwoods semi-quarter-sawn are more likely to be produced since the method is both economic and serves the same purpose; it is done as follows. A breaking cut is first made through the pith of the log; this reduces some of the residual growth stresses and encourages flatter wood. Each half log is in turn placed on the saw table, sawn face down, and then is sawn vertically into boards. The resultant stock, if sold without selection other than for quality, is referred to as 'One Square Edge' as opposed to 'Through and Through', but if a selection is made, and this can often be a cursory one, the stock is now referred to as 'Semi-Quarter-Sawn'.

It will be seen, Fig. 18d that by this method of sawing the growth rings in each board approach the radial direction more nearly than they do tangentially. Reference has already been made to the fact that oak, when sawn midway between true radial and tangential planes frequently produces a more subdued figure and this can hold true for other species. Figure in wood is enhanced by reflected light so any variation in the longitudinal plane from tangential to radial when a large section of timber is converted may well present the structural elements of the wood and its colouring pigments in more attractive form.

Large softwood logs of western hemlock and balsam fir, together with Douglas fir are regularly converted in North America for what is called door stock, i.e. straight-grained planks for use as door stiles and door frames. The trade phraseology covering such stock is '80%

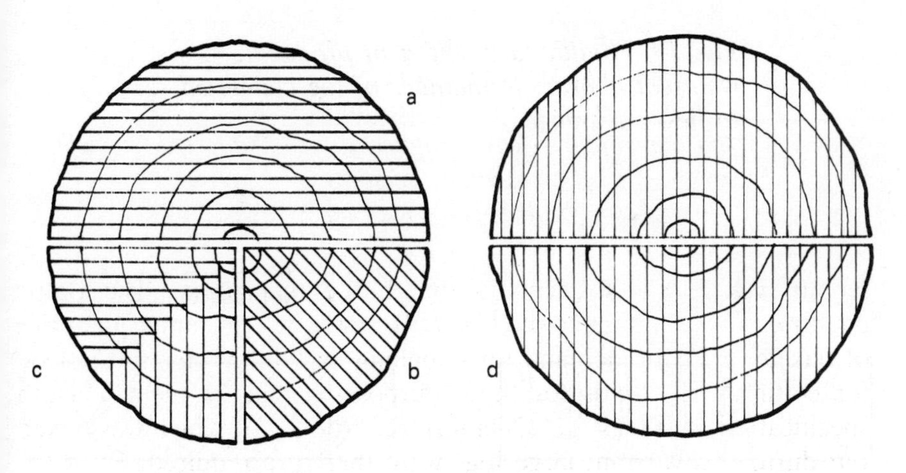

Fig. 18 Methods of conversion
(a) Through and through giving mostly plain sawn boards.
(b) & (c) Methods for obtaining quarter sawn boards: (b) is true quarter sawn
(d) Probably the most practical method. Referred to as "one square edge" it is produced by cutting the log in half and then tangentially sawing at right angles to the 'breaking cut'.

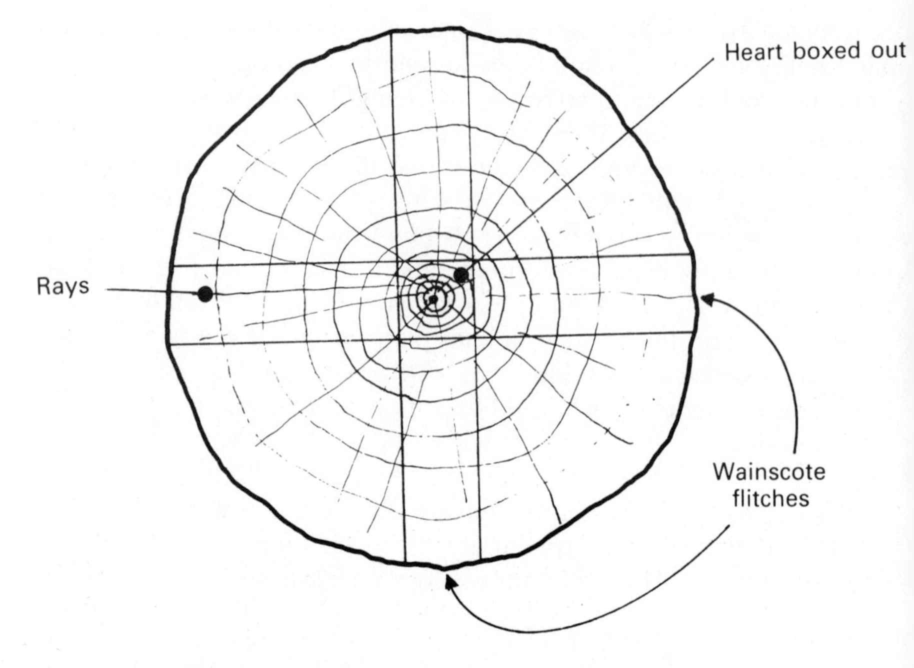

Fig. 19 Traditional method of producing wainscot boards. Remainder of log converted to best advantage

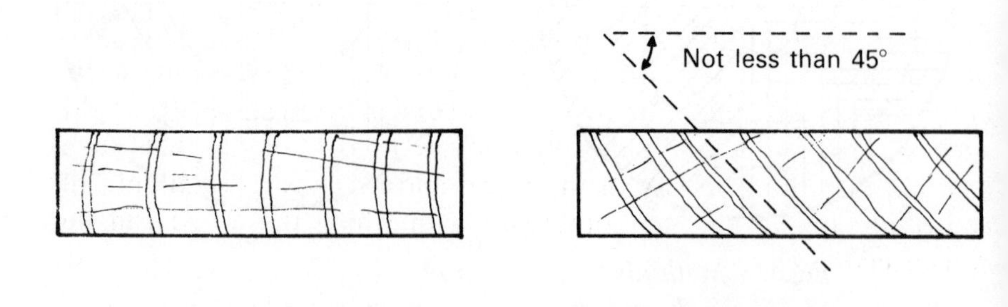

Fig. 20 Edge grain stock need not be true quarter sawn but there are limits if the wood is to shrink less than plain sawn and remain flatter, especially for door frames, flooring and decking

Edge Grain', simply because true radial sawing would be uneconomic, and in this case, unnecessary, because the remaining 20% of any given parcel will not depart very far from the radial.

Factors Affecting Sawing

Some woods, American yellow poplar is a good example, generally accept all forms of cutting without offering excessive resistance to the cutting action or by prematurely blunting the cutting edges; others may prove more difficult. Much depends upon characteristics normally found in a given wood species, or factors abnormal to it. Some abnormalities found in wood are due to the action of parasites, often of fungal origin which, by stimulation of the growing tissue cause localised changes in the woody structure, sometimes creating bird's-eye, typical of some specimens of maple, or, in coniferous trees causing enlarged resin ducts to develop and exaggerating this aspect by increasing the incidence of pitch pockets normally found.

Chemicals such as calcium oxalate and calcium and other lime salts are of frequent occurrence in hardwoods. They are of less importance to dulling of saws in woods such as American mahogany where the deposits are mostly in the cell cavities, than is the case with say, African iroko where calcium carbonate deposits may occur as quite large 'stones'.

Abnormal characteristics, however, generally emphasise the fact that in conversion of baulks or logs of timber there will be a degree of resistance of the wood to the saw, and a degree of blunting to its cutting edge, dependent firstly on the average character of the wood species, and secondly, on the type of saw used. Other factors such as grain direction, density of the wood and its moisture content, all contribute to the ease, or otherwise with which the wood can be converted.

Grain and Texture

The terms, grain and texture are frequently confused and since these have a bearing on the appearance and quality of wood as it falls from the saw it is essential to understand their meaning.

Grain is generally accepted as the direction of fibres determining a plane of cleavage; in other words, a straight-grained piece of wood is one that splits parallel to the axis of the piece, therefore, slanted grain, spiral grain, and cross grain are fairly obvious terms.

Not so obvious are interlocking grain, open grain, coarse grain and fine grain. Interlocking grain is due to a type of growth in which the inclination of fibres is reversed in successive growth layers, a characteristic typical of many tropical woods, e.g. sapele, which shows

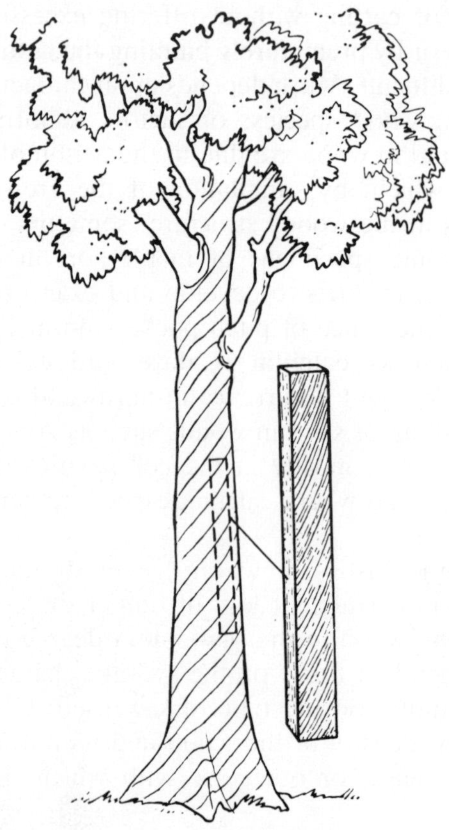

Fig. 21a Spiral growth. Wood from such trees is liable to cause difficulties in conversion and seasoning. This defect can usually be detected in trees where fissures in the bark are in a spiral direction

Fig. 21b Failure due to spiral grain. Note seemingly straight grain within fracture. Breakage has occurred along line of true grain direction. Photograph, courtesy of TRADA

Fig. 21c The edge of the fracture clearly shows true grain direction

a typical ribbon figure when quarter sawn, the light and dark-coloured bands reversing when the wood is reversed because of reflected light from the inclined fibres.

Texture of wood is revealed by touch or by reaction to cutting tools and is largely determined by distribution and size of the various wood elements; thus timbers with a preponderance of large cells will produce coarse-textured wood; those with narrow cells, fine-textured.

The difference between true grain and its direction and the effect of open or close 'grain' is wide since the former is related more directly to end use and the latter to finishing and appearance; a wood with fine texture will probably have a smooth surface as it falls from the saw but it might also have given problems by binding on the saw during conversion because of twisted or spiral grain if this happened to be present.

Where grain direction is critical to a given use, this can be checked firstly by observing any splits, checks or shakes, however small, that have developed in the wood, since they will follow the true grain direction on the longitudinal surfaces; this can be looked for prior to properly cleaning up the wood, i.e. in the raw state or hit and miss finish. The other method is by means of a simple tool which can easily be made up by a local craftsman or by the woodworker himself. The tool, shown in Fig. 23, comprises an L-shaped piece of round metal, the shorter arm of which passes through a wooden handle, the bore or bush being a little wider than the diameter of the metal so that in use the tool moves freely and is not held tight in the handle. The slightly longer arm, projecting almost at right angles to the shaft is arranged to take a sharp, needle-pointed probe, set close to the extremity of the arm. The probe can be of any metal with a sharp point and ideally made to be replaceable by fixing with a grub screw.

To detect grain direction the point of the probe is placed on the surface of the wood and the tool drawn toward the operator, the point scribing a line in the wood in the direction of the fibres. While the average woodworker may feel the use of this technique is unnecessary to his general work, nevertheless there are a good many occasions where, if later problems are to be avoided, excessive slope of grain can amount to a serious fault; the pins of dovetails, for example, can shear off because of cross grain during cutting, with time and wood wasted. But to explain more fully, take another example: a turner is asked to produce some drum sticks, either in ash or hickory, woods ideally suited to this work, and since he has some clean dry wood in hand he goes to work by first ripping out sufficient material. The normal procedure in rip sawing is to first straighten the edge of the board and then rip to width which means the saw is cutting more or less parallel with the axis of the tree; this would be satisfactory for most jobs, but this hypothetical project involves an end use where the product is going to be subjected to impact and shearing forces since a drum stick in use frequently makes

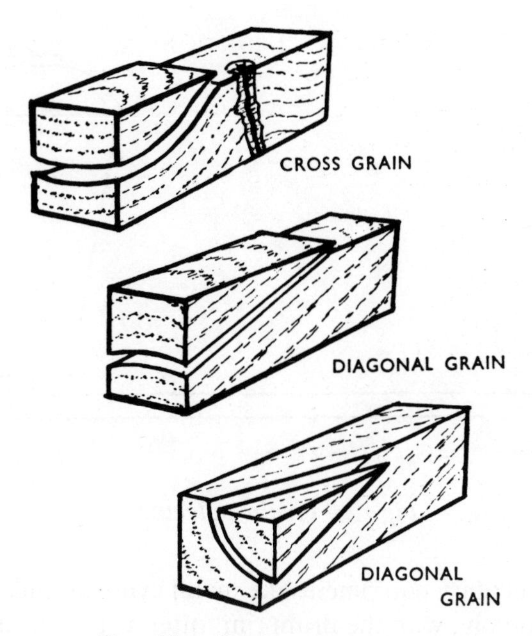

Fig. 22a Grain separation on plain-sawn faces

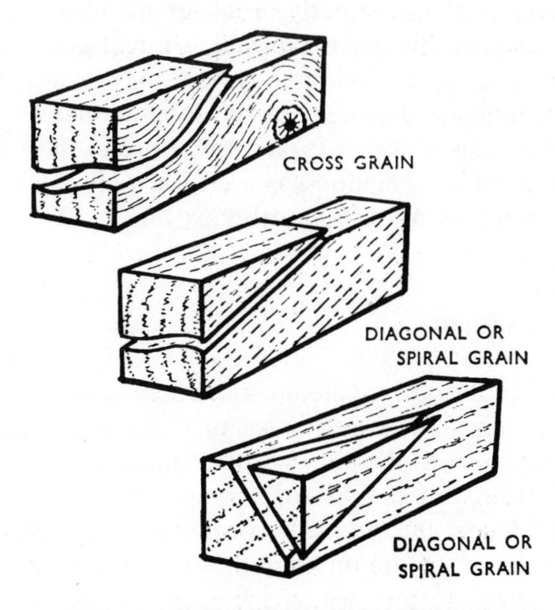

Fig. 22b Excessive slope of grain. Grain separation as seen on quarter-sawn faces

Fig. 23 Grain detector

contact with ancillary equipment like metal cymbals and, intentionally and unintentionally, with the drum rim, often with considerable force. If sloping grain is present in the wood then the main impact, concentrated in the thin neck of the stick, will cause the wood to shear off at an angle quite early in its service life.

Ash and hickory are both typically straight-grained woods as a rule, but any tree, hardwood or softwood, may grow spirally and this results in an oblique alignment of the longitudinal fibres, but not a twisting or bending of these. It pays therefore to have special regard to grain direction if straightness is a critical requirement, even if rip sawing has to follow a more unorthodox line.

Infiltration Content

Various infiltrates and extraneous material found in wood affect its conversion in different ways; silica and oxalates of calcium when abundant contribute to the blunting of cutting edges; copious amounts of gum and resin will slow down sawing; latex canals and in-grown bark require to be cut out and so on. However, the relative ease with which a large section of wood may be converted is governed by a combination of factors which includes density of the wood, its dryness, freedom from large knots, type of grain, and any excess of organic or inorganic matter present. Some materials found in some woods contribute more of a nuisance value to conversion, e.g. tannins,

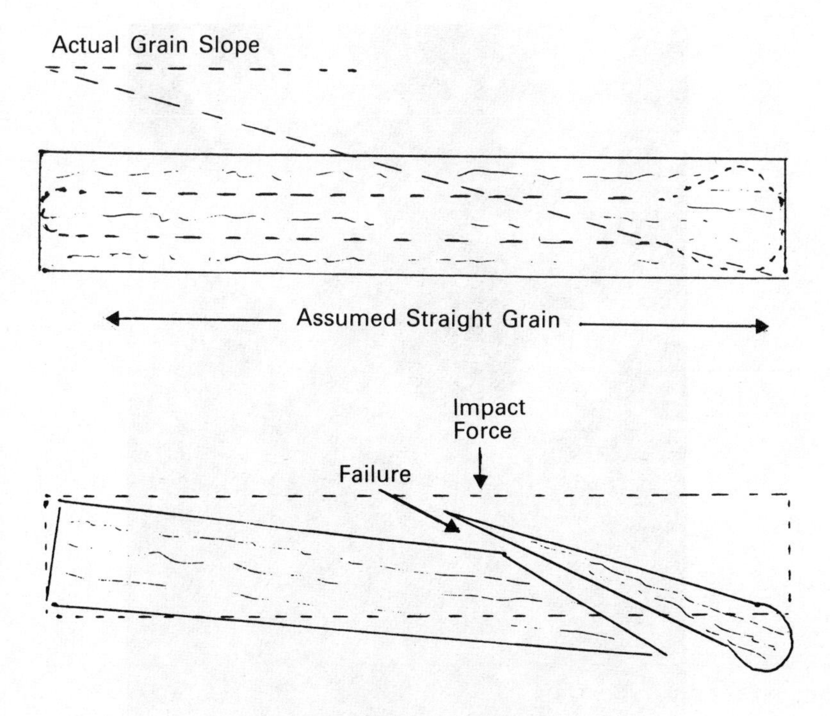

Fig. 24 Excessive slope of grain contributes to impact failure in drumstick

resins and gum, although in this latter respect there are exceptions.

Tannins are astringent substances secreted in the bark and wood of many plants; from the point of view of conversion and storage of timber their ability to react with iron salts in the presence of moisture to produce a black colour means that contact with iron in the shape of machine tables and tools can results in unsightly staining of both the wood and the metal, while storage of tanniferous woods under iron sheeting should be avoided if rain water can drip on the wood. Timber species such as chestnut, oak, and western red cedar are particularly prone to iron staining.

Resins are secreted in glands and canals and where splits and shakes occur in living conifers these often become filled with liquid or solid resin and form the so-called pitch pockets.

True gums are carbohydrate substances and occur in ducts, cysts, and horizontal and vertical canals in hardwoods. There are differences in the composition of resin and gum which should be noted. Resins

Fig. 25 Shagback hickory, George Washington National Forest, Virginia. Photograph, courtesy U.S. Forest Service

are insoluble in water, but soluble in alcohol, whereas the true gums are soluble in warm water, but not in alcohol. Accordingly, gumming up of saws should be regarded as loose phraseology; a sticky residue left on a machine or saw blade after cutting, say, agba, will respond to cleaning with warm water but will be unsatisfactory with white spirit, while the reverse is the case with softwoods like larch, pitch pine and Douglas fir.

Factors Affecting Appearance

The aesthetic appeal of wood results from various causes. Gummy material may interfere with conversion and finishing, but in some instances its presence may brighten the appearance of the wood's

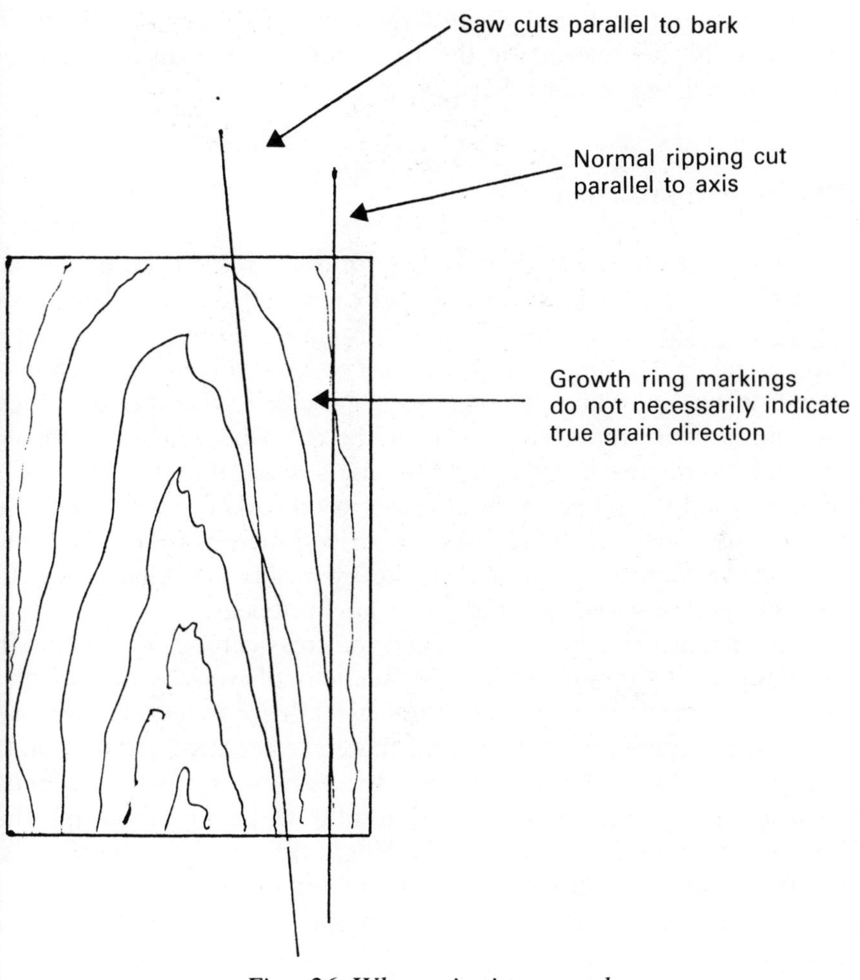

Fig. 26 When ripping wood from well-tapered trees sawing may need to follow unorthodox lines; sawing parallel to the bark is one method

surface by increasing its natural lustre, African walnut being a case in point. However, figure or decoration is the basis of beauty in wood and the principal factors contributing to figure include growth rings, conspicuous rays, bands of soft tissue, and undulating or distorted longitudinal elements. Ornamentation is greatly enhanced by the plane of the section in which the wood is exposed, hence the need to consider a tree before felling and a log before conversion.

Fibre alignment is altered where large limbs of a tree enter the main stem and is the reason for the beautiful crotch figure sometimes found in mahogany and other species.

Chainsaw Conversion

The longitudinal conversion of logs into boards by means of an adapted chainsaw is both feasible and economic and there are times when such a method is the only way to remove timber from the site. Occasionally, a standing tree is offered in a restricted area such as a domestic garden, where it would be impossible to haul out the felled log, but in board form the wood could be removed. Quite apart from this situation is the fact that longitudinal sawing of logs or flitches is often beyond the scope of the average woodworker, and indeed, the farmer and small woodland owner. Provided there is room to fell the tree, then it can be converted on site by means of a chainsaw mill, and the method is not as crude as it might appear.

The standard chainsaw is designed for cross-cutting, i.e. the teeth are designed to remove very small sections of wood which are the ends of the fibres. In rip-sawing, the saw teeth are required to remove longitudinal elements of wood and therefore require modification if they are to do their job properly. In cross-cutting with a normal chainsaw the power employed is intermittent but in rip-sawing the power is continuous throughout the length of cut so the engine must be one which will deliver maximum sustained power, something of the order of 6000 RPM, with a chain speed of say, 3000 feet per minute.

The cross-cut chain can be used for rip-sawing but is not as efficient as one modified for the purpose and these are available from specialist suppliers. The chainsaw mill attachment enables boards to be sawn accurately to thickness, by one person using the smaller version, or by two, with the larger models.

An excellent one on the UK market is the Forestor chainsaw mill attachment which will take logs of up to 32 inches (813mm) in diameter, with a larger version able to saw logs up to 48 inches (965mm) in diameter. The former type is simply attached to the complete chainsaw, while the latter model uses only the chainsaw power head. It can also use two power heads if a faster cutting speed is required.

Both models are fitted with three widely spaced rollers which keep the unit stable while sawing, thus ensuring a clean, straight cut. The first cut is made with the rollers running on a guide plank laid level on top of the log. Subsequent cuts are made with the rollers running on the sawn surface, supporting the chainsaw bar at the desired depth of cut.

Forestor ripping chains consist of one set of specially hardened grooving cutters and one set of central clearing teeth, a combination which provides fast and smooth cutting, causes the least amount of waste, and needs less feed pressure than when a conventional chain is used. These special ripping chains are supplied as follows,

Pitch	*Gauge*
0.375″	0.050″
0.404″	0.058″
0.404″	0.063″

Fig. 27 Forestor Chainsaw mill in action. Photograph, courtesy Forest & Sawmill Equipments (Engineers) Ltd.

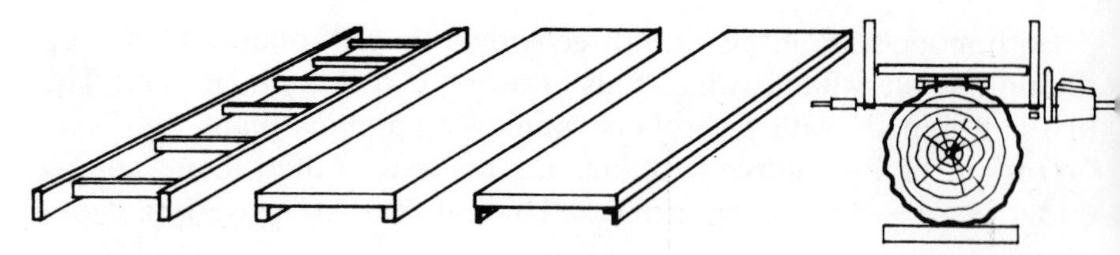

The first cut made is the most important and requires an accurate guide system on which the roller can run. As shown in the illustration this guide can be easily made in one of several ways. It must be secured and supported where necessary to ensure a level cut. It should also project at least six inches beyond the ends of the log to ensure level entry and exit.

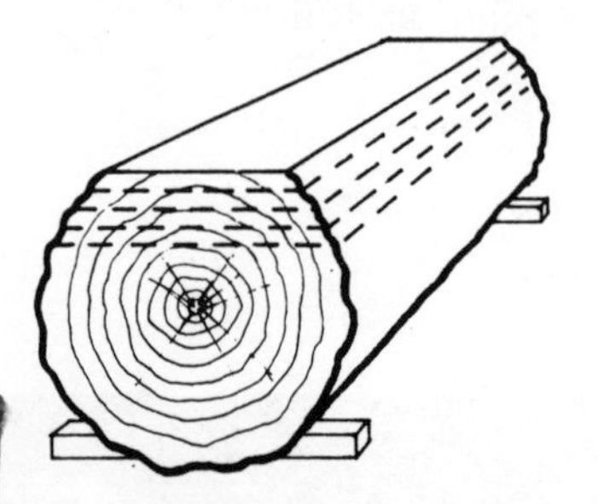

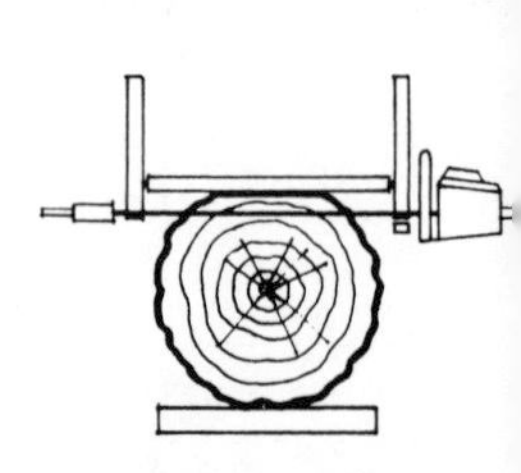

Having completed the first cut the guide plank and the first sawn flitch are removed. The sawn surface is now used as the guide on which the roller run. If the log is to be sawn through and through the depth of cut is set and the second cut is made. Subsequent cuts, varying in depth if required, are then made until the log is sawn.

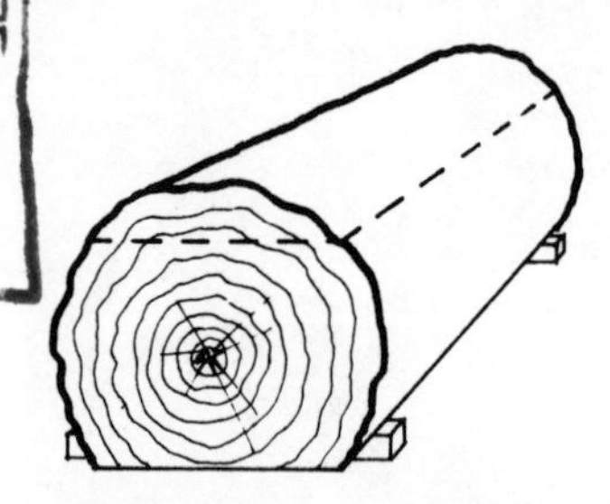

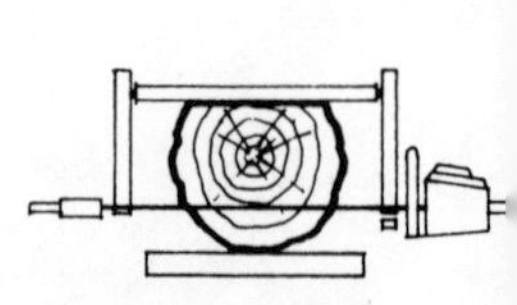

If however, planks of uniform width with straight edges are required, after the first cut the second cut should be made parallel to it, on the opposite side of the log. If the log has a small enough diameter this second cut can be made by simply setting the depth-of-cut deep enough and using the first cut to support the rollers. If the log diameter is too large to do this the log must be turned through 180° and the second cut made with the aid of the guide plank, taking care to ensure that the second cut is parallel to the first.

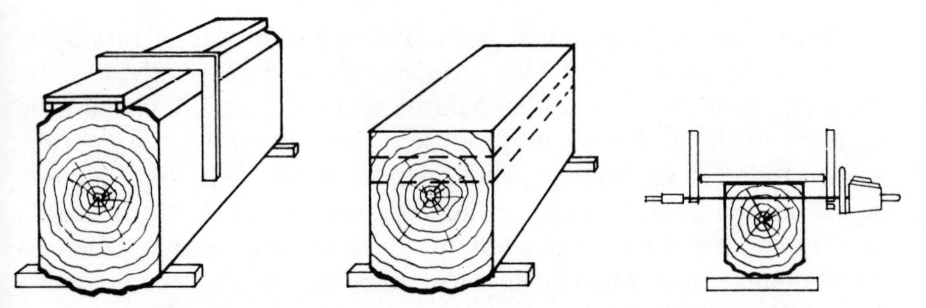

The log can now be turned through 90° and the guide plank resecured. Check with a square to ensure that the third cut will be at right angles to the two faces already sawn. When the third cut has been taken and the rough sided flitch removed the log is ready to convert into planks of the desired thickness, with a uniform width and straight edges.

By standing the planks on edge and clamping them together they can be resawn to produce posts, beams, etc.

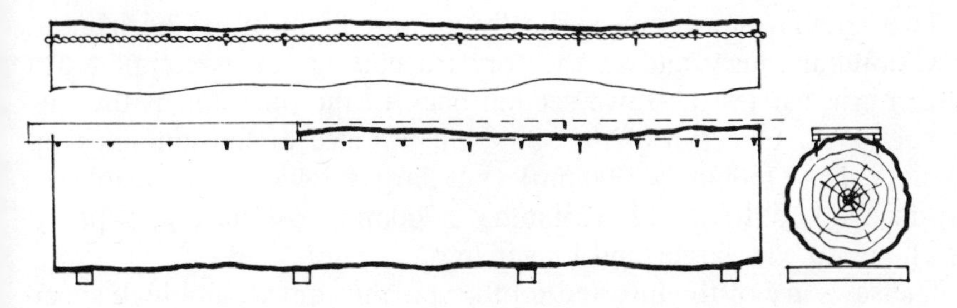

If the log to be sawn is longer than the guide plank, then stretch a rope from one end of the log to the other, at the height at which the bottom edges of the guide plank should sit. Drive spikes into the log so that their tops are level with the rope. This will ensure that the guide plank remains level when moved along the log.

Fig. 28 How to convert logs into sawn timber using the Forestor chainsaw mill attachment

The use of such equipment is an uncomplicated way of producing timber from all types of log, economically and efficiently. As the illustrations show, a certain amount of resawing to width can also be carried out. Some logs could produce nicely figured wood if quarter-sawn, and if large enough could be sawn first into four quarter segments and then, by wedging the bark side to hold the flitch steady, the ripping procedure could now produce suitable boards. There is an advantage, too, when dealing with mis-shapen logs since these can be sawn as they are, without having to cross-cut to shorter, straighter lengths, the produced wood showing distinct grain variation which probably could be put to good use. Free-hand ripping of largish logs into quarters is not easy, but it can be done with the additional use of wedges hammered into the cut to ease the drag on the saw.

Modern Sophistication

Timber conversion has been made more straight forward with the development of sophisticated equipment, particularly in regard to portable machines. Illustrated is the Forestor 2-Cut Mill; a machine which can easily be transported to site and can be set up by two men in about 30 minutes, on even or uneven ground, and without lifting equipment. Logs up to 6′–6″ (2m) in diameter and 18 ft (5.5m) long can be handled and the machine is so arranged as to saw dimension stock direct from the log in one operation.

Two circular saws, one vertically mounted, the other horizontal, saw simultaneously and as the forward cutting run is completed a return arm carries the sawn section back to the operator as the saw travels back. One run of the saws will produce dimension stock up to 6″ x 12″ (150mm x 300mm), but larger saws are available to produce up to 10″ × 12″ (250mm × 300mm), as there are special machines to saw larger and longer logs.

The saws are of the inserted tooth type and tipped, and high speed steel teeth are available. The power source is a 4 cylinder Perkins diesel engine, fitted with special mountings, sawdust screen, etc. This is an ideal saw for estate work or other situations where on-site operations are important to production.

The horizontal bandmill, Fig. 30, again has special advantages since firstly, when mounted on its special self-lowering trailer it is fully mobile for towing to any selected site where it can operate off a diesel or petrol engine and be independent of external power supply, and secondly, it operates at ground level, making handling much

Fig. 29 Forestor 2-Cut Mill. Photograph, courtesy of Forest & Sawmill Equipment (Engineers) Ltd

Fig. 30 Forestor horizontal bandmill. Photograph, courtesy of Forest & Sawmill Equipments (Engineers) Ltd

easier. Log capacity is 5 ft (1.5m) diameter and two or more logs can be sawn side by side provided their joint widths do not exceed the maximum permitted. The length of log to be sawn is limited only by the length of track employed. It will be noted that since it is the bandmill which travels, the log being held on static bearers, the usual log carriage is eliminated, thus reducing capital cost.

A vertical table bandsaw is a versatile machine capable of fast feed speeds and, in modern versions, giving better cutting, less blade problems and enabling larger sections of timber to be sawn. In the model illustrated, Fig 31, the standard machine is a hand-fed table bandsaw with 35.5 inch (900mm) bandwheels and a 4 inch (100mm) blade. It can be used for ripping, re-sawing, and pointing posts.

A power roller feed can be bolted on to turn the standard machine into a production re-saw, with a total of nine feed speeds ranging from 2 to 30 m/minute. Because of the ease of setting up, small runs

Fig. 31 Vertical table bandsaw. Photograph, courtesy of Forest & Sawmill Equipments (Engineers) Ltd

Fig. 32 Hydraulic power roller feed attachment. Photograph, courtesy of Forest & Sawmill Equipment (Engineers) Ltd

are no problem. The power roller feeds are available in electric or hydraulic types, the latter as shown in Fig 32. A further addition to the standard machine is a rolling table fitting which enables small logs to be converted and resawn. The machine is not confined to indoor operation and can be set up on any firm level ground on site.

End Grain Discs and Blocks

A growing tree develops complicated stresses within its bole. Stresses occur in the direction of the growth and create tension near the bark. In turn, compression stress develops in the core of the bole, increasing towards the pith and becoming greater as the tree grows in diameter. When a felled log is converted by sawing lengthwise, especially when the first cut is made through the pith, a lot of stress is relieved, although some residual stress across the growth rings remains and tends to combine with drying stresses formed during seasoning to attempt to pull the wood out of shape, or for it to split.

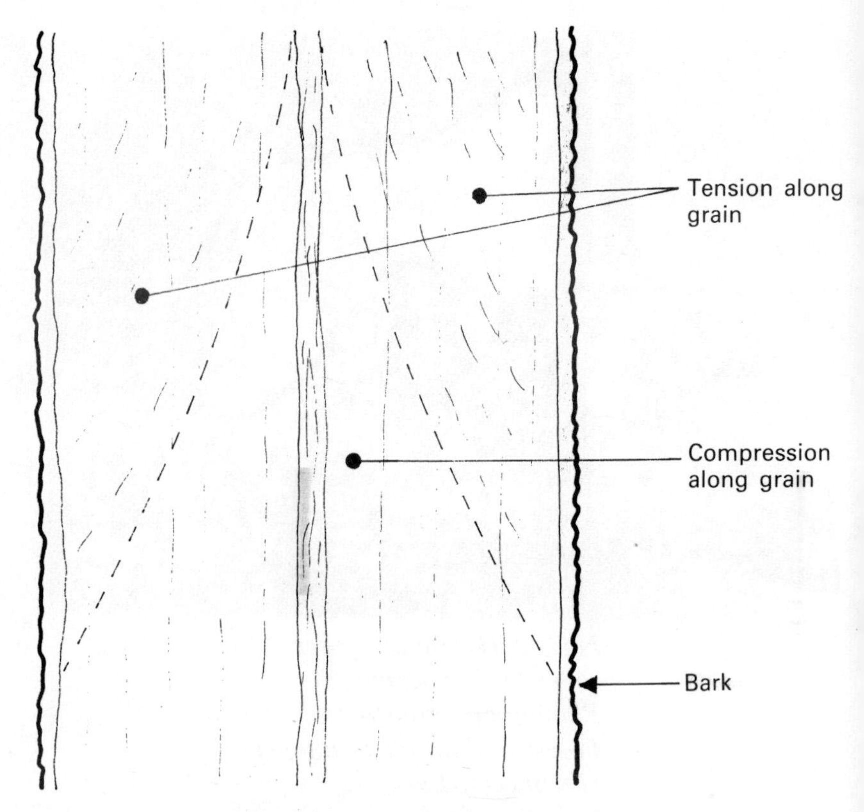

Fig. 33 Growth stresses. Primary stresses develop in the direction of the growth along the grain. Tension is created near the bark thus causing compression in the core.

A felled log, however, often presents an opportunity for slabs or discs to be sawn from its ends, and in this case, because of the relatively short length and wide expanse of end grain, tangential tension forces are triggered off and the wood is encouraged to split badly. Such discs may be treated with polyethylene glycol, as is described later, but much can be done by sawing the wet discs or blocks at an angle, although this does tend to increase the amount of wood used.

Wooden name plates for houses are invariably sawn at angles which primarily give a larger area for lettering, but these are seldom seen with bad checks or splits simply because more long grain is offered by the method of sawing. When a disc or block is sawn square off

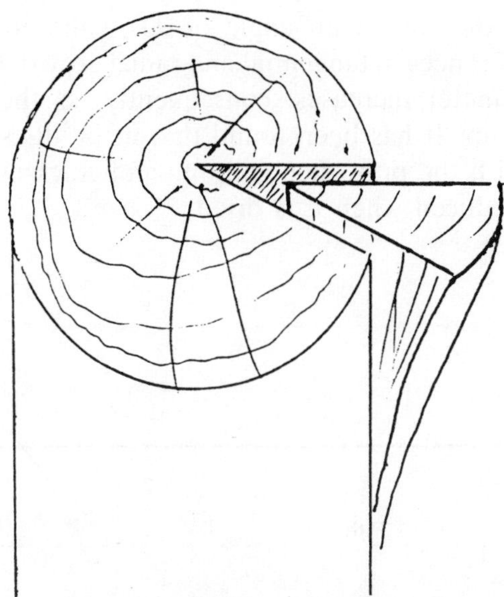

Fig. 34 Effect of relieving longitudinal growth stresses. A wedge shape sawn as above will tend to spring outwards

Fig. 35 Residual stress combines with drying stress and thus encourages splitting of wood discs

the end of a log, i.e. at an angle of 90° with the pith, the force exerted by differences in tangential and radial growth becomes greater as the log diameter increases; consequently, as the wood dries, it tends to open up. It has been found that if the discs are sawn at an angle of 70° with the pith, then splitting and cracking of the wood is considerably reduced when it is dried.

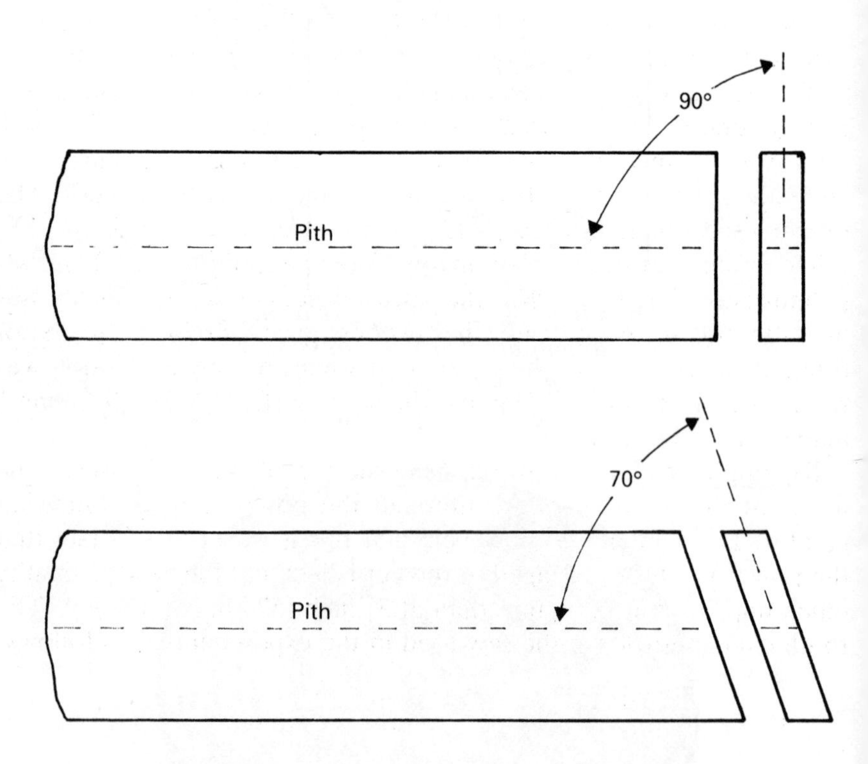

Fig. 36 Reducing the cutting angle reduces the tendency for turnery discs to split during drying.

Rip-Sawn Surfaces

Less force is required to saw wood along the grain than at right angles as in cross-cutting, but quite often the surface quality of the rip-sawn face is much rougher than is desirable. This can result in the off-saw dimensions being tight for finishing to size, especially where the wood is required for edge to edge jointing since more shooting of the sawn edges becomes necessary. Much depends upon the type and character of the wood being sawn, its density and moisture content playing a part. Some woods tend always to being woolly, poplar is an example, but the saw type used has a direct bearing on sawn surface-quality.

In experiments carried out at the Eastern Forest Products Laboratory in Ottawa, it was demonstrated that in rip-sawing, if the saw table was raised so as to reduce the amount of saw protrusion, sawn-surface smoothness approached that of a planed surface. The saw used was a standard 12 inch diameter blade, 0.083 inch thick, with 36 teeth, spring-set for softwoods, and swage-set for hardwoods. The timber used was kiln dried to 12 per cent mc.

The main part of the tests involved the saw protruding 2¼ inches (57mm)above the table when the sawn surfaces of various thicknesses of different softwoods and hardwoods gave varying degrees of roughness. It was found however, that when the saw protrusion was reduced to 13/16 inches (20mm), the sawn surface of ¾ inch (19mm) maple was extremely smooth.

Ripping or flat sawing wood using the top of the blade means the force on the teeth is lower, although the power used is somewhat greater. It is contended, however, that this is offset by the fact that the power per tooth engaged in the cut is less and the surface quality much improved at minimum than at 2¼ inch (57mm) protrusion. The tooth characteristics of the saw used in the experiments is as follows:

Characteristic	**Softwoods**	**Hardwoods**
Set	Spring-set	Swage-set
Amount of set (each side) (inch)	0.02″	0.025″
Hook angle (deg.)	30°	20°
Hook face bevel (deg.)	10°	–
Clearance angle	20°	15°
Clearance face bevel (deg.)	10°	–

The above tests were restricted in the sense that maple appears to be the only species reported upon in the thinner thickness. This is a wood with a fine texture which in theory at least, ought to fall from the saw without too rough a surface. However, it is a wood which, under normal circumstances, tends to vibrate during sawing while the wood may contain curly grain, both characteristics likely to give saw-marked and roughened surfaces. Accordingly, it would seem the results arising from the tests are worthy of further trial.

Saw types are produced by manufacturers to suit the cutting of various types and conditions of timber and are designed according to long experience and data resulting from research. Woodworkers who purchase the smaller type of machine should recognise these are not minor pieces of equipment but scaled down versions of more sophisticated industrial machines. Circular saws supplied with small saw benches and universal machines might have characteristics deemed suitable for sawing all types of wood commensurate with the reduced rim speeds applicable, but other types of saw can be supplied to suit different contingencies on reference to the supplier.

Reverting to the Canadian experiments and the type of saw that was used it is perhaps interesting to note that in the saw types experimented with at the Princes Risborough Laboratory, a similar saw was recommended for use for the rip-sawing of abrasive type hardwoods such as teak and Queensland walnut, as follows:

Saw* type	No.of teeth	Pitch (inch)	Hook (seasoned material)	Clearance	Top bevel	Depth of gullet (inch)
A	32	0.0982 × dia.	15/20°	15°	10°	0.35 × pitch

*Spring-set for rip-sawing at a rim speed of 9000–10,000 ft/min.

TABLE 1: WORKING QUALITIES AND GRAIN CHARACTERISTICS OF SOME EUROPEAN HARDWOODS

Timber	Resistance to cutting	Blunting effect	Grain common to species
Alder	low	mild	fairly straight
Apple	high	moderate	straight to irregular
Ash	medium	moderate	straight
Beech	medium	moderate	straight
Birch	medium	moderate	fairly straight
Box	high	moderate	fairly straight to irregular
Cherry	medium	moderate	variable from straight to cross
Chestnut, horse	medium	moderate	spiral
Chestnut, sweet	medium	mild	straight to spiral
Elm, common	medium	moderate	irregular to cross
Elm, common (Eur.)	medium	moderate	tends to be straighter
Elm, wych	medium	moderate	tends to be straighter
Hawthorn	high	moderate	spiral
Holly	high	moderate	irregular
Hornbeam	high	moderate	irregular
Laburnam	medium	moderate	interlocked
Lime	low	mild	straight
Maple, field	medium	moderate	curly or wavy
Oak	medium (variable)	moderate (variable)	straight to cross
Pear	medium	moderate	straight to irregular
Plane	medium	moderate	straight
Poplar	medium	moderate	straight to curly or wavy
Robinia (false acacia)	medium (variable)	moderate	staight
Sycamore	medium	moderate	straight to curly or wavy
Walnut	medium	moderate	wavy
Willow	low	mild	straight

* See Footnote Table 3

TABLE 2: WORKING QUALITIES AND GRAIN CHARACTERISTICS OF SOME NORTH AMERICAN HARDWOODS*

Timber	Resistance to cutting	Blunting effect	Grain common to species
Alder, red	low	mild	fairly straight
Ash, tough	medium (variable)	moderate	straight
Ash, soft	medium	moderate	straight
Basswood	low	mild	straight
Beech	high	moderate	fairly straight
Birch, paper	medium	moderate	fairly straight to curly
Birch, yellow	medium	moderate	straight
Cherry	medium	moderate	straight
Chestnut	medium	mild	straight
Cottonwood	medium	moderate	straight
Dogwood	high	moderate	fairly straight
Elm, rock & white	medium	moderate	straight to interlocked
Gum, red	medium	moderate	irregular
Hackberry	medium	moderate	irregular, occasionally straight
Hickory	medium (variable)	moderate to severe	straight, occasionally irregular
Hop hornbeam	high	severe	irregular to cross
Magnolia	medium	moderate	straight
Maple, rock	high	moderate	straight
Maple, soft	medium	moderate	straight
Maple, Pacific	medium	moderate	straight
Oak, red & white	medium (variable)	moderate (variable)	generally straight
Persimmon	high	fairly severe	straight to irregular
Poplar, yellow	low	mild	straight
Tupelo	medium	moderate	interlocked to twisted
Walnut, black	medium	moderate	straight

* See Footnote Table 3

TABLE 3: WORKING QUALITIES AND GRAIN CHARACTERISTICS OF SOME AUSTRALIAN HARDWOODS *

Timber	Resistance to cutting	Blunting effect	Grain common to species
Beech, silver	medium	moderate	straight
Black bean	high	moderate	straight to slightly interlocked
Blackbutt	high	fairly severe	straight to interlocked or wavy
Blackwood	medium	moderate	straight, occasionally interlocked or wavy
Blue gum	medium	moderate	interlocked
Brush box	high	fairly severe	interlocked
Cedar, red	medium	moderate	straight
Coachwood	medium	moderate	straight
Gum, red river	high	fairly severe	interlocked
Gum, spotted	high	fairly severe	slightly interlocked
Ironbark	very high	fairly severe	interlocked
Jarrah	high	moderate	fairly straight to wavy or interlocked
Karri	high	fairly severe	fairly straight to wavy or interlocked
Myrtle, Tasmanian	medium	moderate	straight to slightly interlocked or wavy
She-oak	high	fairly severe	straight
Silky oak	low	mild	straight
Stringybark	high (variable)	severe (variable)	straight to interlocked
Tallowwood	medium	moderate	interlocked
Tasmanian oak	medium	moderate	straight
Tawa	medium	moderate	straight to interlocked
Walnut, Queensland	high	very severe	interlocked and and often wavy
White ash	medium	moderate	straight
White beech	medium	moderate	straight

* Based on timber with a moisture content of 10 to 14 per cent. Drier wood will tend to offer greater resistance to cutting with increased dulling effect on cutting edges; conversely, green material will machine with less resistance although generally the dulling effect will be almost the same as for dry wood.

Noise and Woodworking Machinery

Noise, in the context of using woodworking machinery, falls into two categories: one, noise to which operatives may be exposed, and two neighbourhood noise. In the United Kingdom, Codes of Practice exist setting out means of reducing noise in woodworking and sawmilling plants, while the Noise Abatement Act 1960 makes recommendations regarding the statutory position relating to environmental aspects of industrial and some other types of noise.

With regard to the environmental aspect, there is a duty on the woodworker not to impose unnecessary noise on his neighbours and steps should therefore be taken to use the best practical means to minimise the emission of sound arising from woodworking activities.

In very approximate terms, a noise hazard probably exists if it is necessary to shout in order to be audible to a person at 3 feet (1m) distance. In the case of exposure to noise, the use of ear muffs will provide at least 40 decibels reduction at the critical frequency levels of 1000 to 2000 cycles per second (Hz), and in the case of noise giving rise to neighbourhood complaint, then noise containment, by means of insulation or complete or partial boxing-in of the noise source is the only practical means of abatement.

In saw mills, the general consensus of opinion is that any operative exposed to a continuous sound level or its equivalent of more than 90 decibels spread over an eight hour working day should be regarded as requiring ear protection. The non-industrial user of one or two small machines will probably be exposed to no more than 70 decibels at most, but the use of ear muffs or defenders is still important if the machinery is working regularly and the noise causes discomfort, since this could lead to impaired hearing.

One of the problems the woodworker may face is that of having to reduce the nuisance level of noise emanating from his work area, especially during the summer months when more doors and windows are left open, or from the basic construction of the workshop. Complaints from neighbours might lead to legal action being taken if nothing is done to alleviate the noise and it is therefore essential in these circumstances to try to minimise this to a more acceptable level.

Vibrations due to machinery can cause floors to vibrate and create rattling of loose windows, doors and other constructions. These disturbances can annoy workers and reduce their efficiency, quite apart from the noise being transmitted externally. Machines should rest on resilient supports that 'give' somewhat under load; the popular notion that a sheet of cork under a machine will control vibrations is misleading. For low frequency machines, the supports should be damped; for high frequencies, the supports should be elastic without damping.

Sound proofing involves the use of special constructions that reduce the transmission of noise. This may involve the boxing in of certain machines in large plants, but the insulating of walls and ceilings by considering them as partitions is generally more appropriate for small workshops. If one considers that greater insulation than 40 decibels is necessary to reduce loud piano and radio sounds from one side of a partition to the other, some idea can be gained of the degree of insulation necessary for even a small workshop with machines.

Table 4 gives a guide to weight and sound reduction of certain materials that might be used, either to insulate an existing wall or partition, or to erect suitable structures. Manufacturers of acoustic sheet materials supply literature giving results of tests carried out on their products and suitable reference should be made to these when tackling sound insulation problems.

Where a complaint of noise is made, it is wise to seek specialist advice, (a) to check the intensity of noise level both inside and outside the building, (b) to suggest the best method of treatment, and (c) to obtain a written assurance after remedial treatment to the effect that the best practical means have been carried out to abate the nuisance. The phrase, 'best practical means', represents a defence that would probably be acceptable in settling a dispute.

Suitable authorities to approach for advice on noise problems include the following:

United Kingdom: Timber Research and Development Association, Hughenden Valley, High Wycombe, Bucks.

Building Research Establishment, Princes Risborough Laboratory, Aylesbury, Bucks.

Australia: CSIRO Division of Building Research, P.O. Box 56, Highett 3190, Victoria.

U.S.A: Forest Products Research Laboratory, Madison, Wisconsin

Canada: Eastern Forest Products Laboratory, Department of the Environment, Montreal Road, Ottawa, Ontario.

Western Forest Products Laboratory, Department of the Environment, 6620 N.W. Marine Drive, Vancouver 8, B.C.

TABLE 4: SOUND REDUCTION EFFECTS OF VARIOUS MATERIALS USED FOR PARTITIONING (APPROXIMATE FREQUENCY RANGE 1000 c/s)

Description	**Approximate weight (excluding studding of framework) kg/m^2**	**Approximate sound reduction (Decibel)**
Wood, T & G boarding 22mm thick on 75mm × 50mm studs at 450mm centres, boards tightly cramped	Single Skin 13.6	21
12.5mm Fibreboard on wood frame	3.9	24
12.5mm Asbestos insulation board on wood frame	9	25
9.5mm Plasterboard on 100mm × 50mm wood studs at 400mm centres	9	28
Low density particleboard (9 kg/m^2) faced both sides with 3.5mm hardboard, on wood frame		
32mm skin	17.6	26
36mm skin	19.5	27
42mm skin	20.5	28
46mm skin	23.2	30
51mm skin	24.8	31
58mm skin	28	32
50mm low density particleboard, Unfaced.	9	34
Low density particleboard (9 kg/m^2) faced both sides with 9.5mm plasterboard on wood frame		
44mm skin	29	33
54mm skin	33	35
56mm skin	34.2	37

Continued

TABLE 4: Continued

Description	Approximate weight (excluding studding of framework) kg/m²	Approximate sound reduction (Decibel)
36mm low density particleboard (9 kg/m²) on both sides of single wood studs — 38mm cavity	Double Skin 28	33
6.5mm plywood on both sides of 63mm × 38mm wood studs at 400mm centres: two layers of glass silk quilt in cavity	16.2	37
Low density particleboard (9kg/m²) on both sides of staggered wood studs. 75mm cavity with mineral fibre blanket		
22mm skins	18	38 to 39
26mm skins	21	39 to 41
30mm skins	24.4	41 to 43
36mm skins	28	43 to 45

2
Moisture in Wood

There are a number of reasons why wood needs to be seasoned. By reducing its moisture content shrinkage takes place during drying rather than during manufacture and use; gluing and finishing properties are improved; weight is reduced and strength increased; electrical resistance and thermal insulating properties are increased, and the development of stains and mould is resisted.

The amount of watery sap in freshly sawn wood varies according to species and its distribution between sapwood and heartwood is not the same, the sapwood of some species holding a higher proportion of moisture than the heartwood. The main fact is that fresh sawn wood contains an enormous amount of unwanted liquid. A typical beech tree immediately after conversion into timber has a moisture content in the region of 80 per cent and weighs around 60 lbs per cubic foot (960kg/m^3). After seasoning to 12 per cent mc its weight is now 45 lbs per cubic foot (720 kg/m^3); in other words its weight has been reduced by 15 lbs for each cubic foot of its volume.

A gallon of water weighs 10 lbs, therefore this loss of weight means that to dry green beech to a level suitable for use indoors, 1.5 gallons of water must be evaporated from each cubic foot of its volume. This is a conservative estimate since many tree species contain more water on average. Had the example been oak, the loss of weight would have been 17 lbs, not 15 lbs. That aspect, however, is only part of the story since other factors apply, as we shall see. It is sufficient at this stage merely to say that beech is a permeable species and gives up and accepts liquids readily, while oak is very much the reverse in this respect.

Moisture Relations

Water occurs naturally in wood since a growing tree is never dry, but the movement of moisture through, and from, a sawn piece of wood

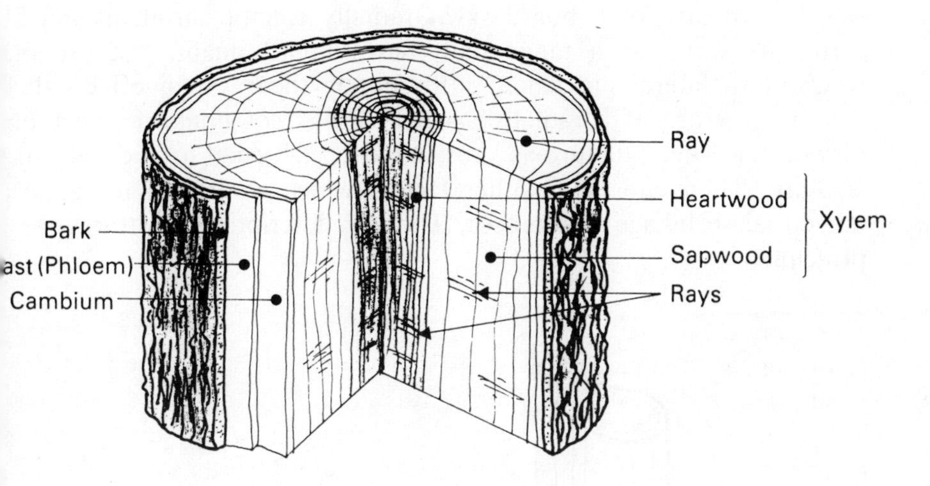

Fig. 37 Sapwood consists of sap conducting tissue and accordingly holds more moisture than heartwood which is supporting tissue. Growth processes initiate in the cambium.

has to be induced by external forces and in such a way as not to disturb, destroy, or distort the woody structure. This poses a variety of problems, none of them insuperable, but many of them subject to human error.

A piece of freshly sawn wood will contain moisture in two forms: free moisture will fill or partially fill the cavities of the cells, much as water is held loosely in a cup, while the cell walls will be saturated with bound, or imbibed moisture.

The molecular structure of the cellulose in the cell walls consists of carbon, oxygen and hydrogen molecules in crystalloid strands or chains. The moisture is held in the cell walls between these strands which are bound end to end. When moisture is removed from the cell walls the molecular strands draw closer together and the wood shrinks, i.e. there is a lateral movement of the wood.

Wood rays were explained in Chapter 1 and it will be fairly obvious that since the ray cells cross the path of the longitudinal cells roughly at right angles, and since the molecular strands in the ray cells are likewise bound end to end and cannot contract in their length, the woody structure of a board sawn radially cannot shrink as much across its width as a tangentially sawn board might, because of restraint to lateral movement afforded by the rays; in effect, the woody structure of a radially sawn board is not unlike a sheet of plywood whose structure is cross-banded in order to increase its stability. There are many other factors which contribute to radial/tangential shrinkage differences. The ray description illustrates the principle.

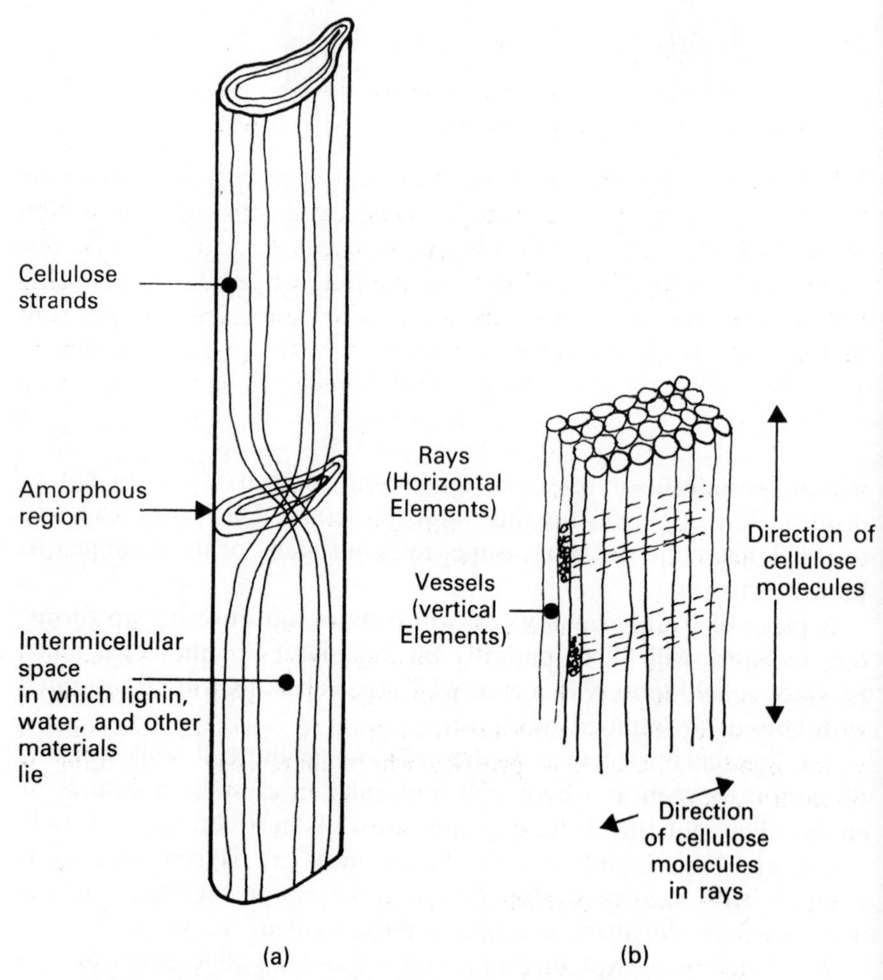

Fig. 38 Vessel elements (a) shows molecular structure. In (b) it will be noted that ray structure restricts movement because the cellulose strands cannot contract lengthwise to a major extent.

When moisture is removed from the cell walls they decrease in their volume and, by contraction, cause a corresponding decrease in the lateral dimensions of the wood, radially and tangentially. The cell cavities remain approximately the same size and the point is mentioned because it is not a narrowing of the wood cells that makes moisture removal relatively slow but other factors, as we shall see.

Fibre Saturation Point

In wood, moisture exists in two forms

i) As liquid in the cell cavities and called 'free moisture'
ii) As 'vapour', or molecular water in the cell walls.

When the free moisture is removed it merely results in a reduction in weight of the piece but a stage is reached, called the fibre saturation point, or FSP, when moisture begins to leave the cell walls. At this point shrinkage commences and will continue while moisture is being lost. Depending on species, this occurs when some 25 to 35 per cent of the original green moisture content has still to be removed by seasoning. For practical reasons, FSP is generally considered to be at 30 per cent mc but it will be appreciated that all parts of a given piece of wood will not reach the FSP stage at the same time; there will be variations in moisture content in a given board and between a pile of boards; consequently, although wood can shrink once the FSP is reached it cannot do so freely in the early stages and stresses therefore develop.

It should be noted that wood, air-dried to say 15 per cent mc has lost about half its total potential shrinkage, while if it is kiln-dried to say 6 per cent mc it has lost four-fifths. The term total potential shrinkage assumes the wood has been dried to zero moisture content which is never the case in practice where there is no such thing as bone-dry wood.

Factors Affecting Moisture Removal

Moisture in wood tends to distribute itself equally throughout the board by moving from areas of high to those of low moisture content which in practical terms means that as the surfaces and ends of a green board discharges moisture into the surrounding air during seasoning moisture from the core of the wood will move toward these

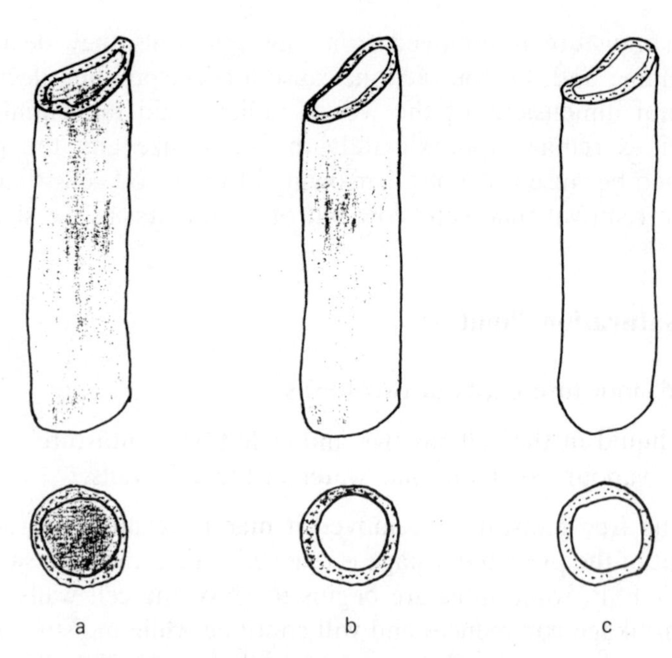

Fig. 39 Vessel Segment
(a) Green wood: walls saturated, cavities filled or partially filled with liquid water
(b) Fibre saturation point: walls wet, cavities mostly empty
(c) Seasoned: walls hold acceptable moisture.

drier areas. As a basic principle this is largely what happens, although it is by no means straight forward.

Although wood has a tube-like structure it is also a complicated one. In the growing state water is conducted vertically through the cells and from one cell to the next by means of openings in the cell walls called pits. The walls consist of several layers of tissue, the centre one, or middle lamella, remaining intact when the pit is formed. In effect, a pit represents a hole through a wall but with an intervening sheet of tissue running through the centre of the wall. Pits are microscopic and in very large numbers in all wood species.

Several types of pit are recognised but it is sufficient for our purpose to mention two, the bordered pit, and simple pit. In the bordered pit, the central membrane is thickened to form a pad called torus while in the simple pit the torus is absent. During tree growth there is a tendency for the torus in bordered pits to become eccentric

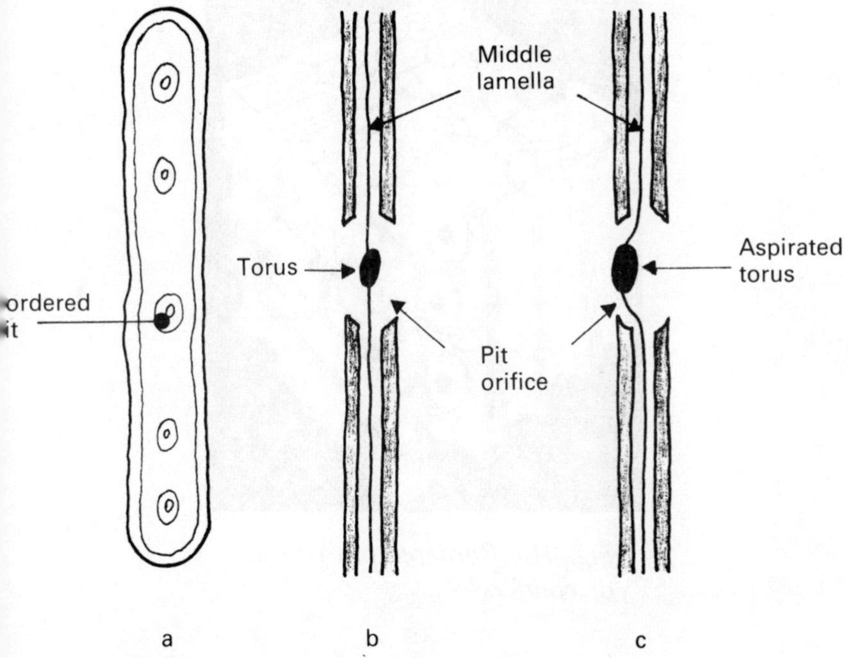

Fig. 40 (a) Softwood tracheid showing bordered pits through which watery sap passes from one vessel to another
(b) Side view of pit with torus in green wood
(c) As wood dries torus tends to be drawn into pit opening thus restricting moisture movement.

and in so doing to block the pit mouth, but this tendency becomes much greater when converted wood is being seasoned.

The technical term used to describe this phenomenon is pit aspiration and there is much evidence to suggest that as moisture is removed from wood during the process of drying, air is drawn through the pit membrane and so displacing it by aspiration. Bordered pits are prominent in all softwoods and they do occur in hardwoods but with much less prominent torus. There is every reason to suppose that the central membrane in simple pitting can similarly be displaced due to aspiration, but it can be seen that when moisture is being withdrawn from a piece of wood its passage is likely to be impeded, firstly, by less permeable tissue, secondly, by gums and resins which may block some of the cells, and thirdly, by growths such as tyloses. These occur in many hardwoods and are due to the fact that while sapwood

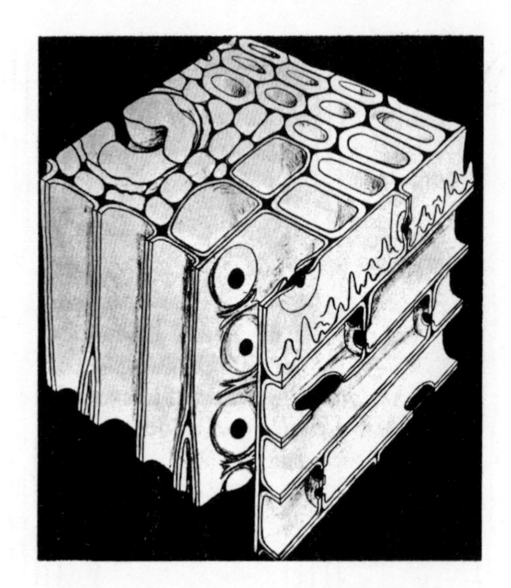

Fig. 41a Bordered pits typical of conifers

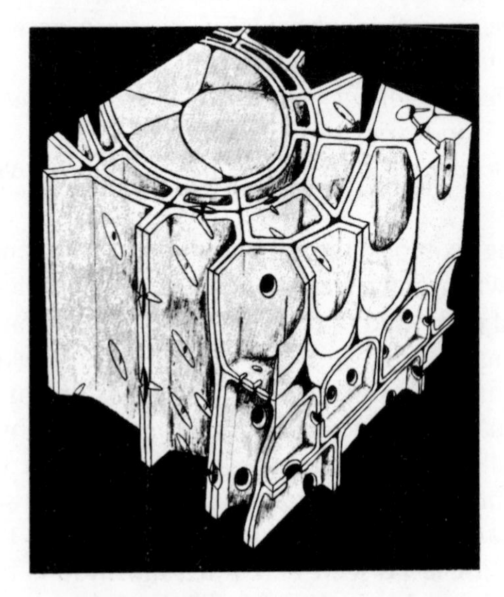

Fig. 41b Simple pits typical of hardwoods. Note Tylose formation in large pore. Not to scale

cells contain water, the adjacent heartwood cells may be empty, the difference in pressure causing the thin pit membrane to grow into the cavity of the vessel cell like a tiny bladder, and since there are innumerable pits the vessel becomes blocked.

It is important to the proper seasoning of wood to understand the significance of tyloses, and probably the best example is to refer to the white oaks whose large, earlywood cells are abundantly blocked with tyloses, thus making the wood ideal for tight cooperage. Red oak cannot be used for the same purpose because in these species the large pores are more open. Because of varying degrees of permeability, some woods are very difficult to dry while others release their moisture more freely.

Directional Flow

During seasoning, moisture moves through the wood in two forms; as a liquid and as vapour. In a longitudinal direction it flows relatively freely but evaporation can be quite rapid from near the ends of the wood. Toward the surfaces of the wood, i.e. laterally, liquid moisture

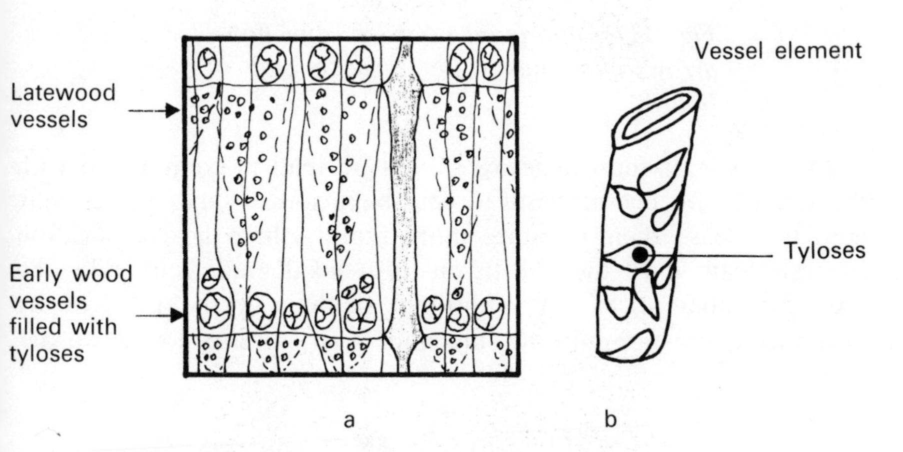

Fig. 42 Tyloses: (a) is a caricature sketch of end grain of white oak showing typical tylose formation (magnification × 8)
(b) Indicates how tyloses impede the flow of liquids through vessels. Middle lamella and primary wall form tiny bladder-like structures.

will be induced by the drying conditions to migrate to these outer zones, but immediately under these surfaces a vapour zone is set up and as long as drying is continuous, a combination of diffusion, transfusion, and capillary attraction will progressively prepare the surfaces for steady evaporation of vapour into the surrounding air. This assumes that the air conditions are satisfactory for steady removal of moisture from the wood at a rate commensurate with its ability to release it and this will vary very considerably with the wood species and its dimensions. Table 5 provides a guide to drying rate of some popular species in open air.

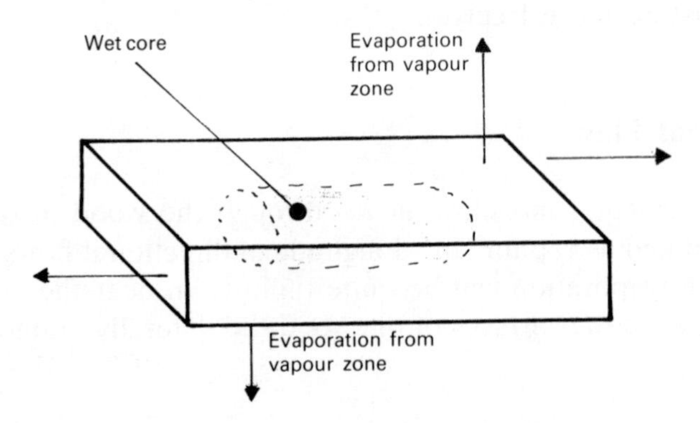

Fig. 43 Evaporation occurs in direction of arrows and tends to be greater at ends of boards.

The slower drying categories should be read in conjunction with the remarks on characteristic drying behaviour. Some species may tend to release their moisture more rapidly than the classification suggests; care must then be taken to avoid degrade, either by the use of thinner sticks, by additional shading of the pile, or by introducing moisture into the surrounding air so as to slow down the drying rate.

Moisture Content Assessment

The standard method of assessing the moisture content of wood is by weight, and although requiring relatively simple equipment to carry out the procedure this is generally of the laboratory type and therefore expensive, at least to the small craftsmen who may only want to carry out periodic moisture content checks. Accordingly, we will discuss what the alternatives are after examining the official procedure, which is as follows:

(1) Select a board to be tested and saw off a piece not less than 1 ft (300mm) from one end, this will ensure that end drying does not affect the result.

(2) From the freshly exposed end cut the sample to be assessed. This should be a section about ½ inch (12mm) in length with the grain. It must not contain a knot, pitch pocket, decayed area, in-grown bark or other abnormality. Tiny slivers of wood must be trimmed off.

(3) The section must be weighed immediately, using a laboratory-type balance; the weight, in grammes being recorded, usually by writing this on the sample in pencil. This is called the initial or wet weight.

(4) The section should now be placed in a small oven, at a temperature of 212°F (100°C) or thereabout and allowed to dry until the weight remains constant, i.e. when all moisture has been removed. This is important to the result, since inaccuracy will greatly affect the assessment.

(5) When the section is dry, it should be weighed again and recorded. This is called the oven-dry weight. The moisture content is now assessed by a simple calculation as follows.

$$\frac{\text{Initial weight} - \text{Oven dry weight}}{\text{Oven dry weight}} \times 100 = \text{moisture content}$$

An example would be Initial weight 40.50 gr. Oven dry weight 30.60 gr.
Therefore,

$$\frac{40.50 - 30.60}{30.60} \times 100 = 32.3 \text{ per cent moisture content}$$

The need for accuracy in weighing and drying samples is easily demonstrated. Suppose, in the above example the weights were rounded down, i.e. 40 and 30 gr; the result now would be 33.3 per cent mc.

If the question of accuracy is observed, and reasonable allowances made for errors, it is possible for the woodworker with limited resources to adapt laboratory procedure to domestic facilities using kitchen scales for weighing samples and ordinary ovens, tops of radiators or other heating devices for drying the samples. The biggest

TABLE 5: APPROXIMATE TIME REQUIRED TO DRY HARDWOODS OF 1 INCH (25mm) THICKNESS IN OPEN AIR FROM GREEN TO 18—20 PER CENT MOISTURE CONTENT

Very Slow – 9 months to 18 months	
Boxwood, Eur. (*Buxus sempervirens*)	Pronounced tendency to surface check
Oak, Eur. (*Quercus petraea & Q. robur*)	Marked tendency to split and check, particularly in the early stages of drying
Oak, Amer. Southern red (Q. *falcata*)	
Oak, Amer. Southern white (Q. *prinus*)	
Slow — 6 months to 15 months	
Apple (*Malus* spp.)	Tendency to warp and split
Ash, Amer. (*Fraxinus* spp.)	Tendency to distort
Beech, Amer. (*Fagus grandifolia*)	Tends to warp, surface check and end split
Cherry, Amer. (*Prunus serotina*)	
Cherry, Eur. (*Prunus avium*)	Tends to warp
Chestnut, Horse, Eur. (*Aesculus hippocastanum*)	Tends to distort and split
Chestnut, Sweet, Eur. (*Castanea sativa*)	Tends to retain wet patches
Hawthorn, Eur. (*Crataegus monogyna*)	Marked tendency to surface check
Holly, Eur. (*Ilex aquifolium*)	tends to distort
Hornbeam, Eur. (*Carpinus betulus*)	Dries Well
Oak, Northern red Amer. (*Quercus rubra*)	Pronounced tendency to surface check especially in early stages of drying
Oak, Northern white, Amer. (*Q. alba*)	
Oak, Northern white, Amer. (*Q. alba*)	
Pear, Eur. (*Pyrus communis*)	Tends to warp
Robinia, Eur. (*Robinia pseudoacacia*)	Tends to warp badly
Tupelo, Amer. (*Nyssa* spp.)	Prone to warping; tops of piles should be weighted down.
Rather Slow — 5 months to 12 months	
Birch, yellow, Amer. (*Betula alleghaniensis*)	Dries slowly with little degrade
Elm, rock, Amer. (*Ulmus thomasli*)	Tends to check and twist
Hickory, Amer. (*Carya* spp.)	Will dry fairly rapidly, but drying should be slowed to avoid warping due to high shrinkage potential

TABLE 5: Continued

Laburnam, Eur. (*Laburnam anagyroides*)	Liable to split
Walnut, Eur. (*Juglans regia*)	Tendency to check
Walnut, black, Amer. (*Juglans nigra*)	Tendency to check
Fairly Rapid — 4 months to 10 months	
Ash, Eur. (*Fraxinus excelsior*)	Tends to end split and distort under high temperatures
Basswood, Amer, (*Tilia americana*)	Dries well
Beech, Eur. (*Fagus sylvatica*)	Moderately refractory; tends to warp, twist and check; requires care
Birch, Eur. (*Betula* spp.)	Tends to warp
Birch, paper, Amer. (*Betula papyrifera*)	Dries well
Cottonwood (*Populus* spp.)	Tendency to warp and twist.
Elm, common, Eur. (*Ulmus* spp.)	Tendency to distort
Elm, wych, Eur. (*U. glabra*)	Less tendency to distort
Elm, white, Amer. (*U. americana*)	Dries well
Lime, Eur. (*Tilia* spp.)	Dries reasonably well; slight tendency to distort
Magnolia, Amer. (*Magnolia grandiflora*)	Dries well
Maple, field, Eur. (*Acer* spp.)	Dries well
Maple, Amer. (*Acer* spp.)	Dries well
Plane, Eur. (*Platanus hybrida*)	Tends to distort
Poplar, Eur. (*Populus spp.*)	Tendency to retain pockets of moisture
Sycamore, Eur. (*Acer pseudoplatanus*)	Inclined to stain and develop stick marks
Willow, Eur. (*Salix* spp.)	Dries well
Rapid — 3 months to 9 months	
Alder, Eur. (*Alnus glutinosa*)	Dries well
Alder, red, Amer. (*A. rubra*)	Dries well
Hackberry, Amer. (*Celtis occidentalis*)	Dries well
Poplar, yellow, Amer. (*Liriondendron tulipifera*)	Dries well

* The approximate minimum drying time is for boards piled in the open when air conditions are most favourable, say in early spring; piling commenced say, in the autumn will naturally take longer to dry. Times for thicker stock are not pro-rata to those given, especially 2 inch (50mm) or more.

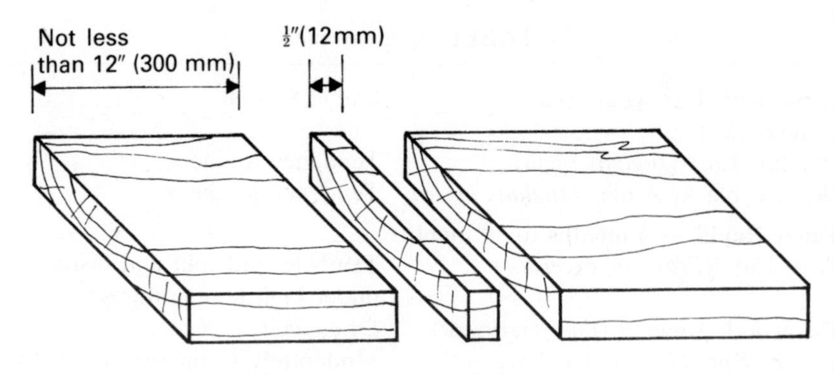

Fig. 44 How to cut sample for moisture content assessment

problem is with accurate weighing; even so, with experience, a very good guide to moisture content can be obtained.

Many kitchen scales today are calibrated in grammes as well as ounces, the latter being more accurate, the former often being rounded off to suit a particular ounce gradation, for example 200 gr equalling 7 ounces. Since the conversion factor ounces to grammes is 28.3495 the equivalent would be 198.4465; a small point perhaps but if domestic equipment is resorted to the margin for error must be reduced as low as possible.

It will be recognised that with the official, professional procedure, time, as well as accuracy is important, hence the generally small size of the samples and the fine measurement of weight. In the example given, hypothetical sample weighed say, 40 gr.wet and 30 gr.dry which in imperial terms equals 1.4 oz and 1.0 oz respectively, hardly enough to give comparison when applied to normal domestic scales. Accordingly, by this method the samples should be larger and more time allowed for drying out.

Drying Progress

In the professional seasoning of wood it is frequently the practice to use the sample board as a gauge to drying progress. If it is possible for the woodworker to have access to larger scales then the following procedure should be followed: after the test sample has been cut the remainder of the board must be weighed and the weight recorded. Before being placed back on the stack the ends of the piece are sealed with a suitable moisture resistant media. By periodically re-weighing the piece it is possible to estimate its moisture content without having to take off further samples; it is done in conjunction with the moisture content already assessed on the small sample.

In the example given, the moisture content was found to be 32.3 per cent and having now weighed the board, this was found to weigh 26 lb and is now the wet weight. The dry weight is found by using the following formula

$$\frac{\text{Wet weight}}{\left(\frac{\text{mc}}{100}+1\right)} = \frac{26}{\left(\frac{32.3}{100}+1\right)} = \frac{26}{1.323} = 19.65 \text{ lbs estimated dry weight}$$

Assume that when the next check is made the weight of the board has fallen to 23.5 lbs; the moisture content then would be:–

$$\left(\frac{\text{Current weight}}{\text{Dry weight}}\right) - 1 \times 100 \left(= \frac{23.5}{19.65}\right) - 1 \times 100 = 20 \text{ per cent mc}$$

This method gives only a guide to drying progress. It is based on the assumption that firstly, the oven dry sample was a true average of the whole of the parcel of wood, and secondly, that the moisture was evenly distributed throughout the consignment which is seldom the case.

Moisture Meters

Where, for one reason or another, it is not possible to check moisture content by standard methods, recourse can be made to a moisture meter; a small, convenient piece of equipment which operates on the principle that wet wood represents an excellent conductor of electricity but dry wood is a very good insulator. The most popular type is based on direct electrical resistance of the wood, and by driving pin-type electrodes into the wood the resistance is measured and translated into moisture content values.

When the pins are inserted into the wood in the direction of the grain, current from a battery flows through the wood between the electrodes and thence back into the meter which is fitted with a calibrated scale and a pointer indicating moisture content. More sophisticated instruments have a digital display together with small, plug-in program keys which automatically set the calibration required for different timbers.

Moisture meters are ideal instruments, particularly for the man with limited resources, but they must be used properly. Generally, they are calibrated by the manufacturer for a certain species of a certain thickness at a certain temperature; accordingly, the manufacturer's recommendations regarding possible correction factors for some timbers must be referred to.

As the moisture content of wood is reduced its electrical resistance nears its extreme and recordings of moisture range from about 6 per cent at the lower level to fibre saturation point of 25 to 30 per cent at the higher level.

To obtain proper readings from a meter however, it is essential the pins are driven into the wood to their full depth and this is not always easy with the denser woods and furthermore, that the depth of penetration of the pins is at least one-quarter of the board's thickness.

When moisture content checks are made on drying wood this always has a core wetter than the surface zones, i.e. it contains a moisture gradient. There are ways of equalising out the contained moisture but only under the controlled conditions of a kiln as will be explained later. Therefore, when a test sample is checked for moisture content by the oven dry method the result is correct only for the test board and represents an average for the parcel itself, but it does not show the distribution of moisture throughout the thickness; where this is required, further test samples must be cut. An appropriate sample is suitably marked in pencil to facilitate reassembly and is then sawn into five, roughly equal size strips; each strip is individually weighed and oven dried and its moisture content established so that when the pieces are reassembled, a moisture distribution pattern can be noted.

The trade, for reasons of economy, usually works on average values of moisture content which means an average of a number of boards based on an average value of their thickness. Suppose for example, a parcel of timber has to be kiln dried to 10 per cent mc without any other qualification, and after drying, five test samples read 12, 9, 10, 11 and 8 per cent respectively. This would produce the 10 per cent mc value requested but it would nevertheless give a fairly wide spread of moisture throughout the parcel of wood as a whole.

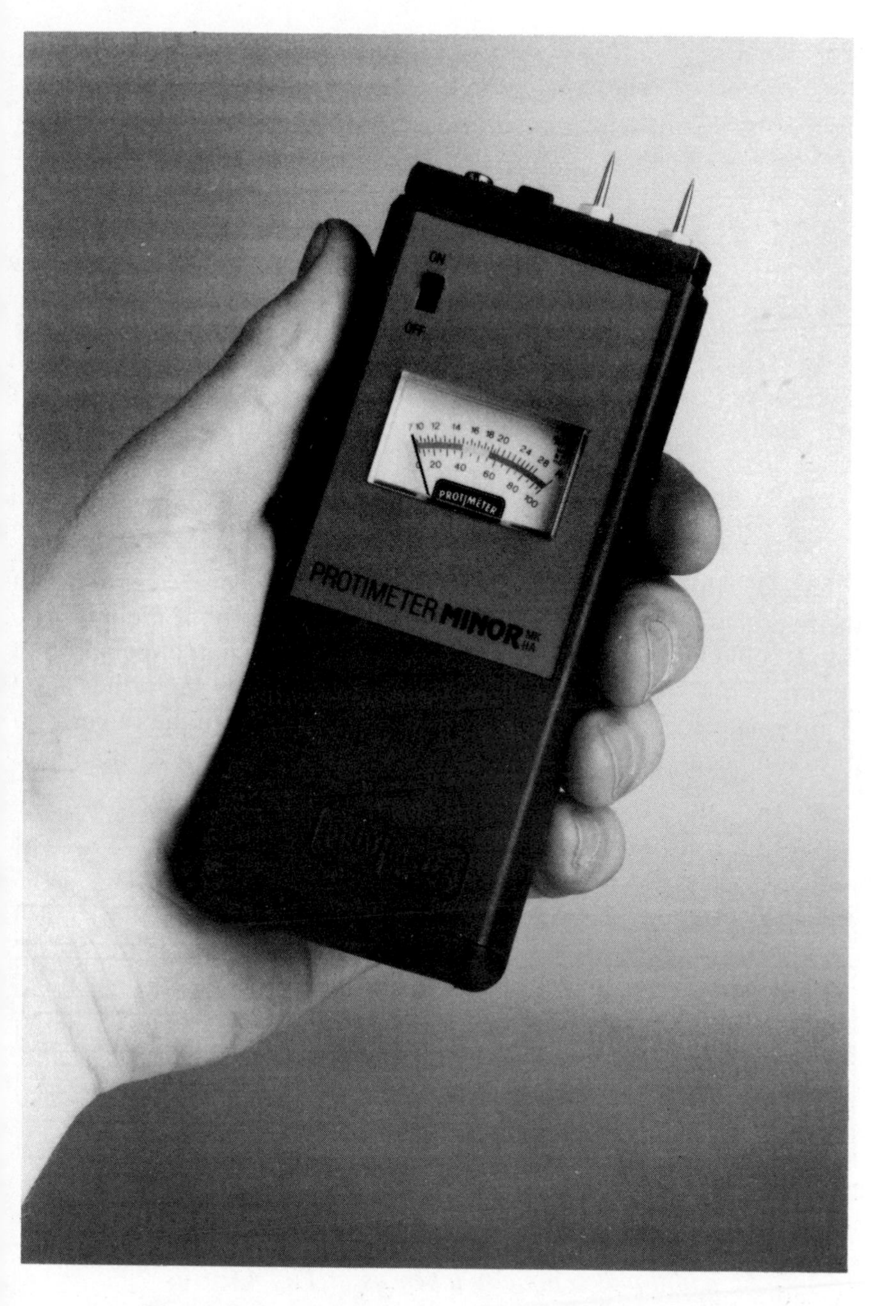

Fig. 45 Small moisture meter ideal for the craftsman. Photograph, courtesy of Protimeter Plc

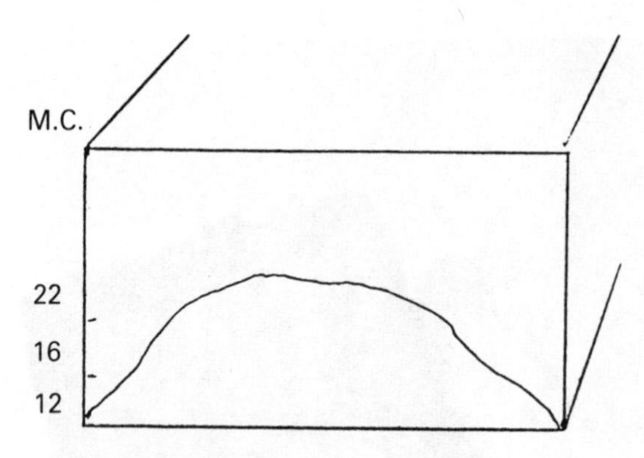

Fig. 46 Moisture Gradient: in practice the variation of moisture content is less uniform than shown here

Largely, this system works quite well, but there are exceptions. In the hypothetical example mentioned, a closer control of the moisture values could be had by including a qualifying clause in the specification or order, e.g "to be kiln dried to 10% ± 2% m.c. when measured in any one piece", but since this would involve longer time to perform it would incur additional expense. Meters used by the trade are

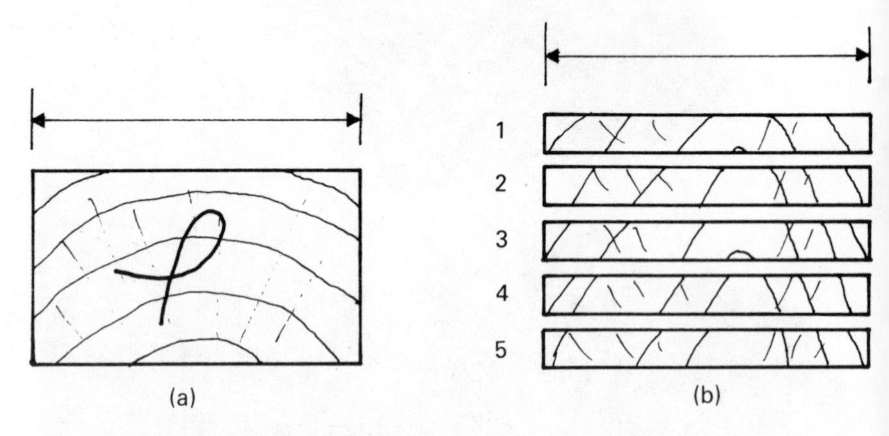

Fig. 47 Moisture distribution test: An additional sample is taken as in Fig. 44. It is then marked (a) to facilitate moisture pattern determination after sawing into five equal pieces (b) they are assessed by moisture meter or by weighing and drying in an oven.

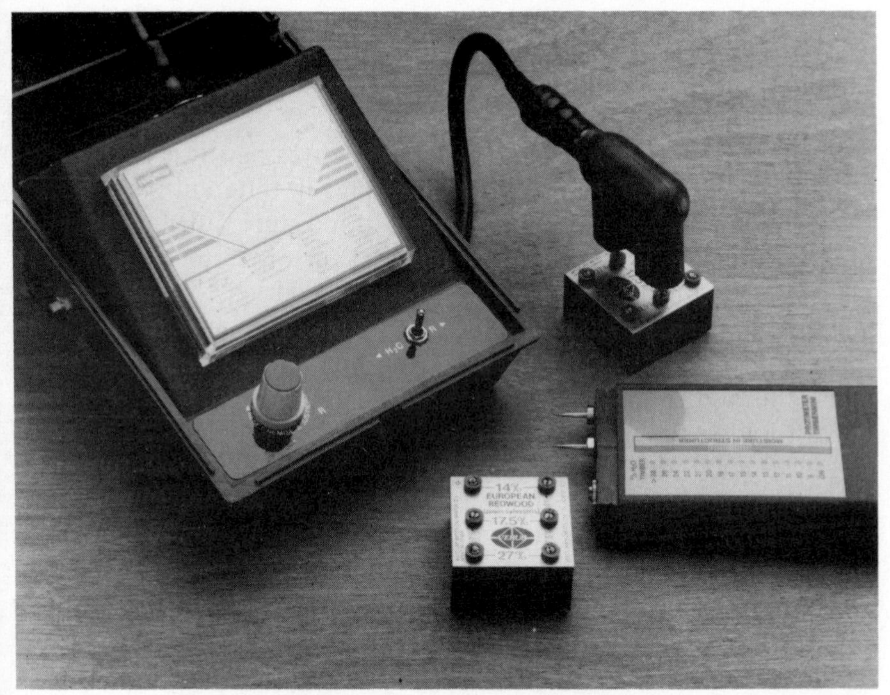

Fig. 47a Checking the accuracy of a Protimeter Timber Master moisture meter

Fig. 47b Verus moisture meter calibrator

supplied with insulated electrode pins which will penetrate the wood to a depth of about 30 mm.

With a moisture meter equipped with shorter pins it is possible to saw boards in half and by inserting the pins in different parts of the thickness to obtain an immediate result of moisture distribution. This is often done, especially with thick stock in order to obtain some idea of how much longer the wood must be dried to a reasonable level at mid thickness. It is not a very economic method however since the wood must be shortened and more than one piece needs to be checked to give a fair result.

With a little thought and the use of a moisture meter it is possible to gain a good idea of how wet the centre thickness of any wood is at a given moment, by a simple calculation based on the assumption that the moisture gradient is a parabolic curve; often it is more sinusoidal than parabolic, but the assumption is not unreasonable.

Calibration

Moisture meters are calibrated, i.e. set to a standard on a particular timber species. Throughout Europe, this is invariably European redwood (*Pinus sylvestris*). Meter manufacturers then supply conversion tables covering other species to which the meter readings are corrected where necessary. In use, a given meter is subjected to extreme conditions which tend to stretch the leads and weaken the electrode heads due to the force with which they are frequently driven into the wood. Consequently, the manufacturers of these precision instruments recommend an annual service. This is not always observed by users, often not until the battery has obviously run down, but by then doubtful readings of moisture content has occurred.

In an effort to provide moisture meter users with on-site calibration checks the Timber Research and Development Association (TRADA) developed a prototype calibrator under the name TRADA Checkbox. This has since been further developed and patented and is now manufactured under licence as the Verus Moisture Meter Calibrator.

This is a very useful little instrument measuring only 40 × 40 × 25mm and weighing only 58 grams. Models are available to fit all leading makes of resistance type meter of UK and European origin. In the experimental work it was found that in meters not operating as intended, margins of error were progressive across the measured range, i.e. from the wet, medium, to dry positions. Accordingly, the

Verus Calibrator was developed to electronically simulate three specimens of a single wood species at three checkpoints, 14%, 17.5% and 27% moisture content at 20°C with an accuracy better than ± 0.2% m.c.

In use, the instrument determines the accuracy of readings on analogue, digital, and light-emitting-diode type moisture meters; precisely checks the low, medium and high points on the moisture meter scales; checks meter leads and electrode heads for breaks in circuit; gives early warning of meter battery failure, and takes only about 20 seconds to carry out a full check.

Meters can be calibrated for all popular species including ash, Brazilian mahogany, Douglas fir, European redwood and whitewood, keruing, meranti, oak, etc. The power requirement is a batteryless passive electrical circuit so there is no battery to fail or replace. Recommendations are that the Verus Calibrator is used before taking readings in order to establish whether or not the meter is accurate and to enable immediate corrective action to be taken if necessary. Corrective action, to cover most contingencies short of a complete breakdown, is detailed in instructions supplied with each Calibrator covering each individual make of meter.

Verus Instruments Ltd, are located at Ash House, Church Lane, Bledlow Ridge, High Wycombe, Buckinghamshire HP14 4AZ.

3
Shrinkage and Movement of Wood

Wood is sensitive to moisture and will respond to atmospheric changes by shrinking and swelling. However, in the practical use of wood the implications should be considered as separate issues, i.e. as direct shrinkage, and as movement by shrinking or swelling. Direct shrinkage varies with the species and with the fibre direction, but movement is rather more predictable.

For most practical purposes the direct shrinkage of wood is considered as being proportional to the amount of moisture lost in drying below a fibre saturation point of about 30 per cent. In order to obtain scientific data on the drying behaviour of different woods it is necessary for tests to be based on a common factor; accordingly some tests call for total shrinkage values. Total shrinkage is based on drying green wood to 0 per cent moisture content, while other tests use a green to kiln dry basis of 12 per cent mc. Generally, British publications give shrinkage values from 30% to 12% moisture content; while U.S.A. publications give values of 30–20% m.c; 30%–6% m.c. and 30%–0% m.c.

Broadly speaking, direct shrinkage values give only a guide to potential loss of volume of timber bought green, and to the small user of wood, once he has paid for his stock, is of relatively less importance than say the safeguarding of the initial quality he has selected; but shrinkage, or the tendency for drying wood to diminish in size, is of extreme importance since it imposes forces which will attempt to seriously spoil the initial quality.

Movement

Probably the most annoying situation in woodworking is to find that movement is occurring in piece parts prior to assembly or, worse still, in finished work. Usually, unwanted shrinkage is the culprit, but wood may swell if the conditions encourage it.

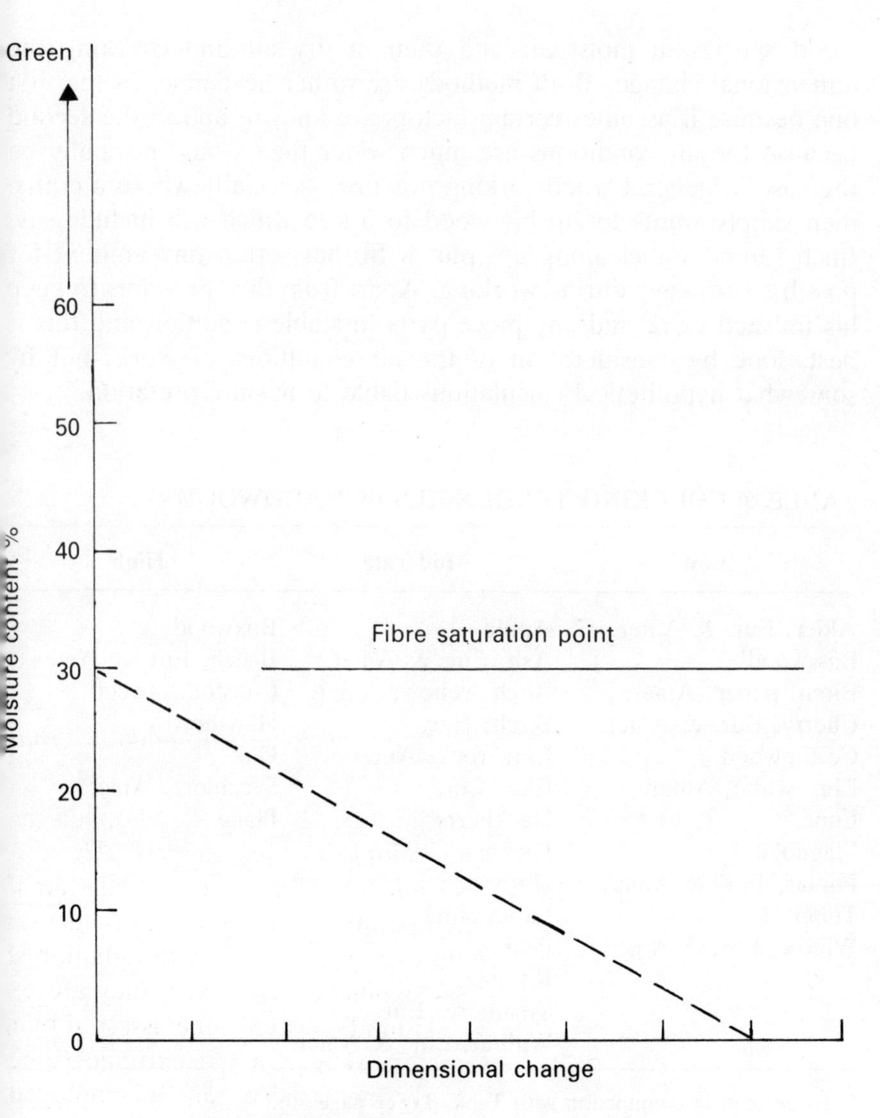

Fig. 48 Moisture content relationship — shrinkage & swelling

Technically, there are two principal ways in which movement by shrinking and swelling can be assessed: one, by calculating the anticipated amount of shrinkage likely to occur between initial and final moisture contents, and two, by conditioning samples of different wood species in moist air and then in dry air and checking the dimensional change. Both methods are rather academic: in the first one because it assumes certain factors are known, and in the second because the air conditions are much wider than would normally be the case in general woodworking practice, especially where a craftsmen simply wants to rip his wood to a size which will include say, $\frac{1}{8}$inch (3mm) for cleaning up, plus a further, often tiny amount for possible shrinkage during working. Apart from this, he wants to keep his finished work and any piece parts in stable condition and this is best done by consideration of the air conditions of work, not by somewhat hypothetical calculations liable to misinterpretation.

TABLE 6: CHECKING TENDENCIES IN HARDWOODS*

Low	Moderate	High
Alder, Eur. & Amer.	Apple	Boxwood
Basswood	Ash, Eur. & Amer.	Beech, Eur. & Amer.
Birch, paper, Amer.	Birch, yellow, Amer.	Chestnut, sweet
Cherry, Eur, & Amer.	Birch, Eur.	Hawthorn
Cottonwood	Elm, rock, Amer.	Oak
Elm, white, Amer.	Elm. Eur.	Sycamore, Amer.
Lime	Hackberry	Plane
Magnolia	Hickory	
Poplar, Eur. & Amer.	Holly	
Tupelo	Laburnam	
Willow, Eur, & Amer.	Pear	
	Robinia	
	Sycamore, Eur.	
	Walnut, Eur. & Amer.	

* To be read in conjunction with Table 5 (See page 70).

However, the following formula is used to convert total shrinkage values into dimensional changes.

$$S = \frac{(MI - MF)\,D}{\left(\frac{30}{ST \text{ or } SR} - 30\right) + MI}$$

S = Shrinkage or swelling in inches or mms.
MI = Initial moisture content per cent
MF = Final moisture content per cent
D = Dimension in inches or mms. at initial moisture content
30 = Fibre saturation point per cent
ST = Total tangential shrinkage in per cent, divided by 100
SR = Total radial shrinkage in per cent divided by 100

To give an example of its use, assume a white oak board, plain sawn, i.e. tangentially, 12 inches (300mm) wide, with a moisture content of 18 per cent is to be used in a situation where it will dry further to 10 per cent mc; what shrinkage is likely across the wide face of the board? Total shrinkage of white oak, from 30% moisture content to 0% moisture content, in a tangential direction is 0.9 per cent.

By substitution in the formula the following is derived,

$$\frac{(18 - 10)\,12}{\left(\frac{30}{0.09} - 30\right) + 18} = \frac{96}{321.3} = 0.299 \text{ inches}$$

The potential shrinkage, or swelling in adverse circumstances would be approximately 0.3 inches. This assumes that relevant factors are known, but to the comparative layman, values of total shrinkage are not always available. Some are published by technical laboratories, but even these do not include most of the odd specimens likely to be acquired by the craftsman; however, we will return to the subject.

Generally, the term shrinkage is used technically, to describe the initial loss of dimension when drying from green (ie Fibre Saturation Point).

Movement describes the loss, or gain, in dimension when the wood, in service, responds to changes in the atmospheric environment.

The method of assessing movement as distinct from shrinkage is as follows: small samples of wood are prepared from stock whose moisture content is less than 15 per cent. These are next conditioned in air at 77°F (25°C) and 90 per cent relative humidity until they attain constant weight. Each sample is then measured accurately on both faces; the sections are then conditioned again in air of the same temperature but with 60 per cent relative humidity, after which they are measured again and the results computed as a percentage of the width.

By this method the corresponding tangential movement for white oak is 2.8 per cent, or 11/32 inches per foot of width; this amount, i.e. 0.344 inches compares closely with 0.3 inches obtained in the previous example; but again, it assumes the woodworker has all the facts and figures to hand. There is a point that should be noted here since it is relevant to this discussion, that initial shrinkage may produce different results than the corresponding movement values.

Direct shrinkage from fibre saturation point down to a moisture content low enough for most woodworkers to want to start work will account for a high proportion of total shrinkage and, depending on species, will vary in its amount to a considerable extent. However, as the moisture content of all species attains equilibrium with the surrounding air conditions of say a warm room or workshop, movement by shrinking, or by swelling, due to air condition variations remains fairly constant.

A very rough guide to potential movement is to consider 0.04 inch per foot of width per each one per cent moisture content change for tangentially sawn, i.e. plain sawn wood, and 0.025 inch per foot for radially sawn or quarter sawn stock. In metric terms this equals 1mm per 300mm for plain sawn and about 0.625mm for quarter sawn boards, the exact conversion factor, millimetres to inches being 0.03937, here called 0.04.

Predicting movement

The potential movement likely to occur in a given set of circumstances can be predicted provided the moisture content of the wood is known and the environmental air conditions understood. Suppose for example, a table top 24 inches (600 mm) wide was to be produced from plain sawn wood in 4 inch (100 mm) widths. The wood is air dry, with a moisture content of say, 15 per cent, but the equilibrium moisture content of the workshop and the living room in which the table will eventually be used is say 10 per cent, a difference of 5 points; accordingly, the potential movement in each piece would be 0.04 inches x 5 = 0.20 inches shrinkage per foot of width, but since the wood is 4 inches wide the potential shrinkage would be 0.20 inches divided by 3 = 0.0666 inches or 1/16 inch per piece.

There are a number of ways of regarding this; the actual amount is relatively small, but if several pieces of wood are placed side by side the shrinkage gap between adjacent pieces is not 1/16 inch but half as much again because of accumulative shrinkage. Again, it might be argued that wider pieces would reduce the joints, which is obvious, but with wider stock there is likely to be a further problem of cupping which not only would disturb the level of the surface but would exaggerate the shrinkage.

Naturally, the example given simply indicates the movement potential and in practice, if the movement were to take place after

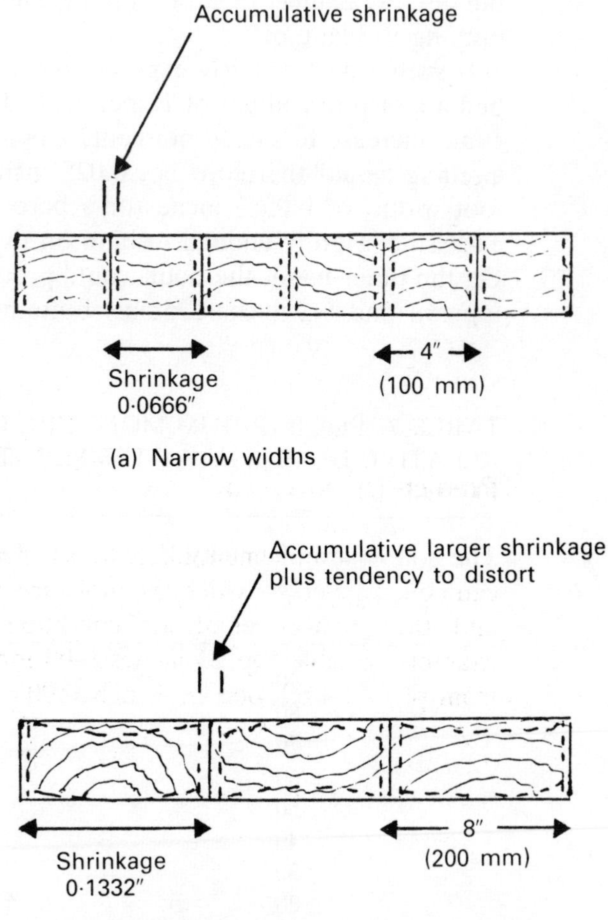

Fig. 49 Shrinkage potential in panel — drying from 15% mc to 10% mc. In practice, method of jointing, edge to edge, loose tongue, or T & G would restrain shrinkage, but shrinkage forces still apply and joints could open up, and in the case of wide sections, could distort.

the top was made up there would be some restraint offered by the method of jointing and by the fixing of the top to the under frame, but the wood must attempt to move under the conditions described, and while the joints might hold firm in part they would not at the ends which would tend to open up. If the top was framed up then the frames would tend to shrink an equivalent amount.

If this is looked at in reverse, i.e. movement by swelling, let us consider an external door whose stiles are, again, 4 inches (100mm) wide, but in this case the wood is edge grain or quarter sawn. Most doors are produced from kiln dried material and produced in suitably warm surroundings and are similarly stored. It is sometimes the practice to 'shoot in' a door in order to make it fit the frame loosely, but let us assume that on this occasion the door fits nicely with nothing to plane off.

If such a door is fairly exposed in use, and at the time of hanging had a moisture content of 12 per cent, during the winter months this could increase to say 19 per cent, 7 points more moisture. Potential swelling would therefore be, 0.025 inches × 7 = 0.175 inches per foot width, or 0.0583 inches or a bare 1/16 inch across the face of each stile. Hardly enough to bind on the frame, generally speaking. On the other hand, the door would probably have a 4 inch (100mm) top rail and a 9 inch (225mm) bottom rail, invariably made from

TABLE 7: EQUILIBRIUM MOISTURE CONTENT RELATED TO RELATIVE HUMIDITY AT TEMPERATURES OF 32°F (0°C) to 100°F (38°C)*

Relative humidity %	**Equilibrium moisture content (approx.)** %
90	19 – 17
80	16 – 15
70	13 – 12.4
60	11 – 10.3
50	10 – 8.9
40	8 – 7.2
30	6 – 5.7
20	4.5 – 4

* EMC becomes lower as temperature is increased at a given relative humidity.

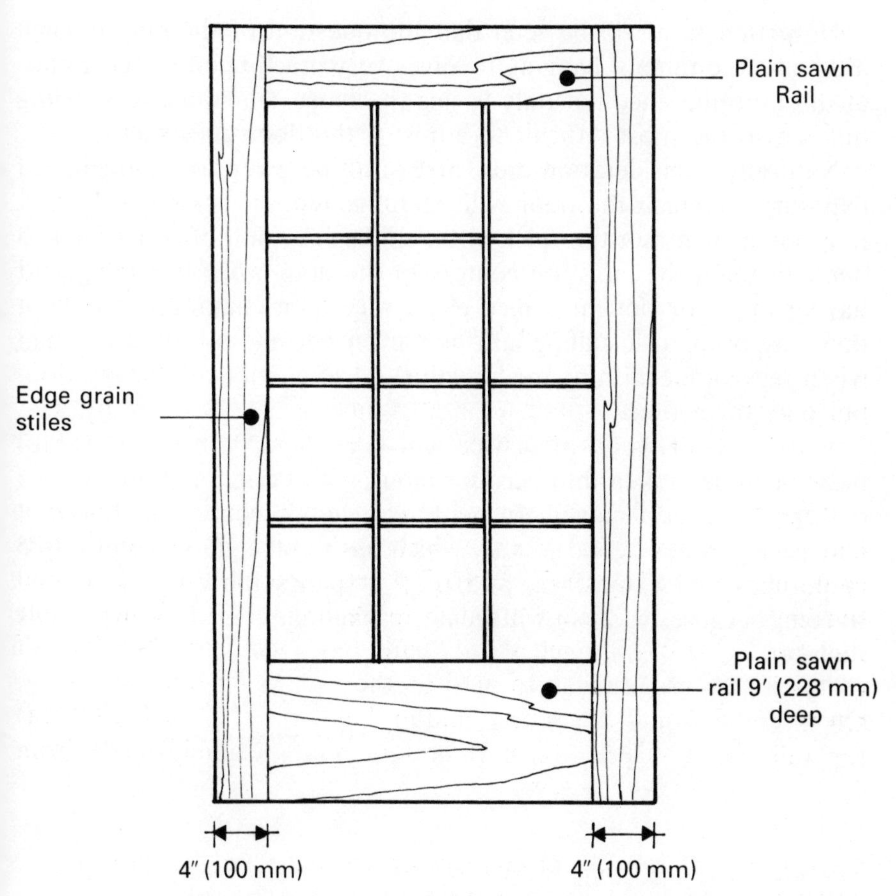

Fig. 50 Potential movement in a door is likely to be greater in the rails, two 4 inch (100mm) stiles radially sawn have less movement potential than one 9 inch (228mm) tangentially sawn rail, since movement in a radial direction is only about five-eighths that tangentially. Doors should be allowed to equalise to room conditions or exposure conditions, before hanging.

plain sawn wood. The probable swelling under the conditions described would be 3/32 inch (2.4mm) across the width of the top rail and 3/16 inch full (4.8mm) across the bottom rail. Much of this swelling will occur above the lower edge of the rail tenon i.e. within the door frame itself.

Nevertheless it will be seen that the question of shooting in such a door at the time of hanging revolves very much upon the clearance at the bottom, since not only is this the most vulnerable to swelling but is also the most difficult to adjust if the door does stick.

Naturally, consideration must first of all be given to the degree of exposure to which the door will be subjected. If it is set well back in a porch then moisture pick up could be less and, of course, much depends upon the moisture content of the door when it is hung, but having made or bought a nice clean, dry door, naturally it is kept that way prior to installing and may even become a little drier than when received. Finishing media will slow down moisture absorption, but only for a time.

Movement of wood in service will occur and if the basic values mentioned are taken into account along with design and application of a product then movement generally can be contained. It is not enough to produce good joinery which has to take glass, for example in porches and extensions, and then fit panes fairly tightly in the frames, because to do so will invite cracked glass if the wood swells slightly.

4
Drying Elements

Drying and seasoning are synonymous terms although liable to be misconstrued. Faults that may have developed in woodwork are often dismissed as being due to the use of unseasoned wood, or timber may be offered as being well-seasoned, but in both cases the term 'seasoned' has no precise meaning; accordingly, it is appropriate to recognise that seasoning is the process of drying timber to a moisture range appropriate to the conditions and purposes for which it is to be used.

If wet, freshly sawn wood is left exposed to the air it will begin to dry, very slowly if the weather is inclement, more quickly if the sun is shining. The speed with which it will dry will also be governed by the dimensions of the boards, and by the amount of air that is in actual contact with the wide faces of the boards, and by the relative permeability of the wood species in its ability to release its moisture.

The essential elements which cause wood to dry naturally, i.e. as in a timber yard, is wind and sun. If we want to speed up the process, the timber is placed in a kiln chamber and the drying elements are now heat, humidity, air circulation and ventilation, the process sometimes being referred to as artificial drying. This is a confusing term since it implies that the resultant dried wood has lost some of its character, a point of view occasionally expressed by woodworkers who contend that wood dried 'naturally' in the open air has a better character in all respects than that dried in a conventional kiln or dehumidifying chamber, since now some of the 'nature' has been removed from the wood.

The contention is erroneous; the character of wood is inherent to the species and while it may be spoilt by poor seasoning practices there is less chance of this happening where there is complete control of the elements and this can only apply to sophisticated equipment such as kiln and dehumidifying chambers.

Quite apart from the fact that artificial drying is much faster than air drying, is the fact that for interior uses, wood cannot be dried to a low enough moisture content by air drying, at least not in temperate climates, and further drying is essential, and the most satisfactory way to achieve this is by kiln drying. Some countries have their own special problems in relation to seasoning their native woods, Australia being a particular case. Many Australian timbers are highly refractory, and do not respond kindly to moisture removal, but specialised techniques have been developed whereby highly stressed wood is reconditioned in the kiln to satisfactory proportions. There are some very beautiful and useful woods in Australia, but they would be much less useful without kiln drying expertise.

Heat

Water is evaporated from wood by the application of heat. The greater the heat the faster the drying. Heat causes moisture to migrate towards the surfaces of the wood and there to be evaporated. Whether in air drying or kilning, the rate of drying is intimately related to the applied temperatures. Because kilns operate at temperatures higher than ambient, kiln drying is faster than air drying.

Normally, there is no intrinsic advantage in drying slowly at low temperatures if the wood will tolerate faster drying at higher temperatures. Obviously, in open air drying there can be no control over ambient temperatures and therefore no prediction of drying rates.

Humidity

Air, at a given temperature, cannot hold more than a definite amount of water in the form of vapour; when it contains that amount, it is described as saturated. Expressed another way, if air is saturated the actual weight of water vapour contained in a given volume is termed the absolute humidity. Thus saturated air at 110°F (43.5°C) has an absolute humidity of 26 grains of moisture per cubic foot (0.03m^3). It is usual to express the amount of water vapour in the air as a percentage of the amount required to saturate the air at that temperature, and this is referred to as relative humidity or RH.

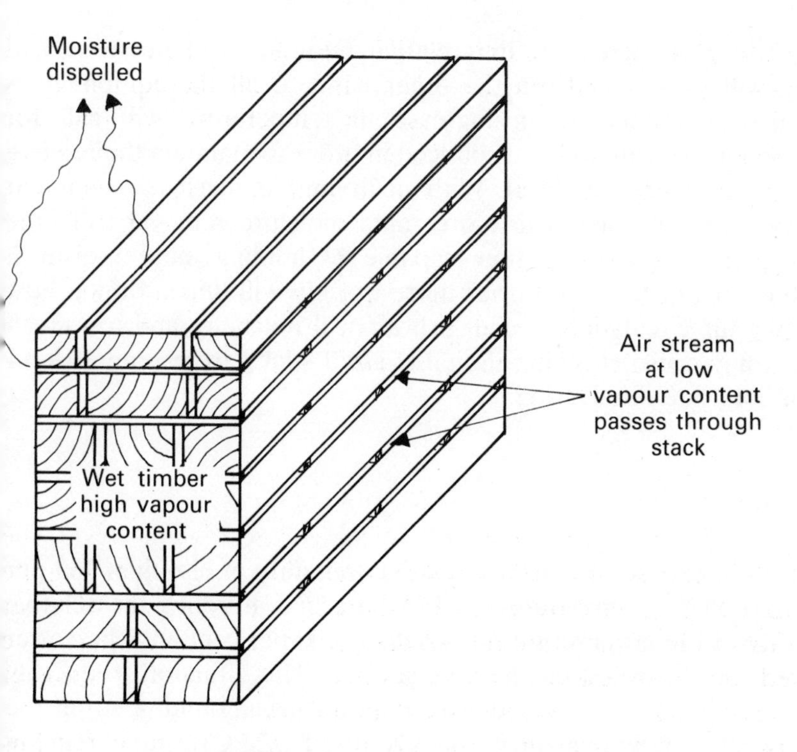

Fig. 51 Moisture in drying wood is attracted by air at low vapour content. This movement of moisture is induced by air flow.

In the example given, i.e. at 110°F one cubic foot of air would hold 26 grains at saturation equalling 100 per cent RH. Suppose the air only contained 13 grains of water at that temperature, then the relative humidity would be 13/26 of 100 or 50 per cent. It will be noted from Table 8 that if the temperature of the air is increased its moisture holding capacity is likewise increased. This is important because increased temperatures may have different effects on wood according to the method of drying. In kiln drying, the combination of heat, air circulation and venting off of moisture laden air operates in step with various stages of relative humidity. Therefore, if the temperature is increased, the relative humidity will decrease and drying will be speeded; on the other hand, if all the equipment is operating as intended, an increase in temperature will call for additional live steam to be introduced in order to maintain the relative humidity at the desired level. With air drying, an increase in ambient temperature will encourage more rapid moisture removal from the wood because the air is now capable of holding more moisture. Whether or not the wood dries more quickly will depend upon how good the air circulation is. Sluggish air will not help, and too much hot air may cause checking. Suitably sized sticks are essential to air circulation. See Air Drying.

Dew-Point

If a surface exposed to air at a relative humidity of less than 100 per cent is cooled, a temperature will be reached at which the relative humidity of the air close to the surface is 100 per cent and thereafter dew will be deposited on the cool surface. This situation frequently occurs in air drying of wood especially during autumn. During the day the air conditions might be say 75° F (24°C) and a relative humidity of 60 per cent, i.e. with 6 grains of water per cubic foot of air. During the night the temperature drops say to 50°F (10°C) and it will be seen from Table 8 that at that temperature the air can only hold 4 grains of water at 100 per cent RH so the excess 2 grains of water per cubic foot is now deposited as droplets or beads of dew.

Measuring Relative Humidity

The measuring of humidity is generally done by using one or other of the many forms of hygrometer available and adapted to meet special requirements. In the field of timber drying the wet and dry bulb hygrometer is in general use. Basically, they consist of two similar thermometers; they may be mercury thermometers or one of the electrical types, i.e. resistance or thermoelectric, but it is more usual for mercury thermometers to be used. The bulb of one is normal, while that of the other is covered with a thin cotton sleeve, wetted with distilled water. Water evaporates from the wet bulb at a rate depending mainly on the relative humidity and rate of air movement. It is found, under proper conditions of air movement, that the relative humidity can be determined simply from the actual air temperature recorded and the difference in temperature between the dry and wet bulbs.

As the air velocity past the wet bulb increases from zero, so the bulb temperature falls. A steady temperature is reached at quite a low air velocity; for example, about 3–5ft/sec for a mercury thermometer. Thereafter the depression of the wet bulb remains constant up to quite high air velocities. It is the practice to obtain the relative humidity from standard tables covering a range of values based on the differences between dry bulb and wet bulb temperatures. Suppose the dry bulb reads 105°F and the wet bulb reads 100°F, i.e. a wet bulb depression of 5 points, then the relative humidity, as read from a standard table or chart is 83 per cent. Table 9 gives a guide to this.

These are delicate, sophisticated instruments used mainly in the professional drying of timber and as such need to be maintained

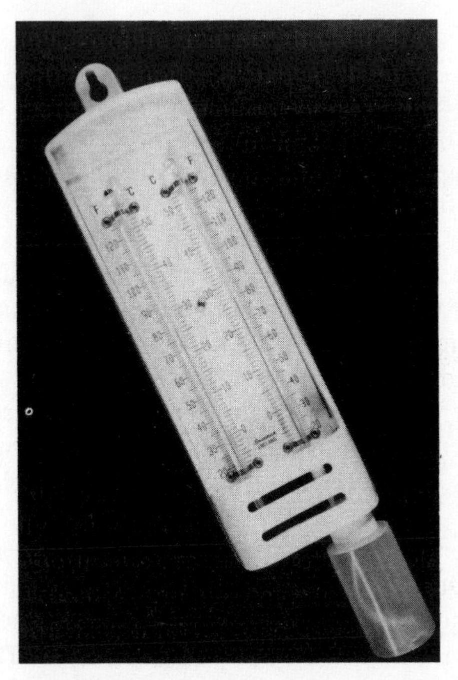

Fig. 52 Wet & Dry Bulb Hygrometer. Photograph, courtesy of Brannan Thermometers Ltd

TABLE 8: RELATIONSHIP BETWEEN TEMPERATURE AND THE MOISTURE HOLDING CAPACITY OF AIR AT SATURATION

Temperature °F	°C	Grains of Moisture per ft^3 1 grain = 1/7000 lb avoirdupois
120	49	35
115	46	30
110	43.5	26
100	38	20
90	32.3	15
80	26.5	12
70	24	10
60	21	8
50	15.5	6
40	10	4
	4.5	3

properly, otherwise sources of error can creep into the readings. A few of the contingencies giving rise to error are:

1. Too low a fan speed may result in insufficient air velocity; this ought to be 3ft/sec or more.
2. Badly fitting or dirty sleeving will tend to give a high wet bulb reading. They should be replaced once they get dirty.
3. When wetting the wet bulb care should be taken not to get water on the radiation shield since this will tend to humidify the air passing through the instrument.
4. Impure water will give a high wet bulb temperature. Distilled water is preferable for wetting but clean drinking water can be used, but this will necessitate more frequent replacement of the sleeving.

There are other methods of checking air humidity and other forms of instrument. One that is frequently used is the sling, or whirling hygrometer, which is simply a frame holding the wet and dry thermometers, with a wooden handle fixed to the frame, somewhat in the form of a rattle sometimes used at sporting events. The thermometers are ventilated by whirling the frame around the handle at a rate of rotation specified by the manufacturers. After about one

TABLE 9: RELATIVE HUMIDITY RELATED TO DRY BULB TEMPERATURE AND WET BULB DEPRESSION

Wet bulb depression °F	**Dry bulb temperature °F**								
	50	60	70	80	90	100	120	150	200
				Relative humidity %					
1	93	94	95	96	96	96	97	98	98
2	86	89	90	91	92	93	94	95	96
3	89	83	86	87	89	89	91	92	94
4	74	78	81	83	85	86	88	90	92
5	68	73	77	79	81	83	85	87	90
6	62	68	72	75	78	80	82	85	88
7	65	63	68	72	74	77	80	82	86
8	50	58	64	68	71	73	77	80	84
9	44	53	59	64	68	70	73	78	82
10	38	48	55	61	65	68	72	76	80
11	32	43	51	57	61	65	69	74	79
12	27	39	48	54	58	62	67	72	77
13	21	34	44	50	55	59	65	70	75
14	16	30	40	47	52	56	62	68	74
15	10	26	36	44	49	54	60	66	72
16	5	21	33	41	47	51	58	64	70
17	–	17	29	38	44	49	55	62	69
18	–	13	25	35	41	46	53	60	67
19	–	9	22	32	39	44	51	58	66
20	–	5	19	29	36	41	49	57	64
21	–	1	15	26	34	39	47	55	63
22	–	–	12	23	31	37	45	53	61
23	–	–	9	20	29	35	43	51	60
24	–	–	6	18	26	33	41	49	58
25	–	–	3	15	24	30	40	48	57
26	–	–	–	12	22	28	38	46	55
27	–	–	–	10	19	26	36	45	54
28	–	–	–	7	17	24	34	43	53
29	–	–	–	5	15	22	33	42	52
30	–	–	–	3	13	21	31	41	51

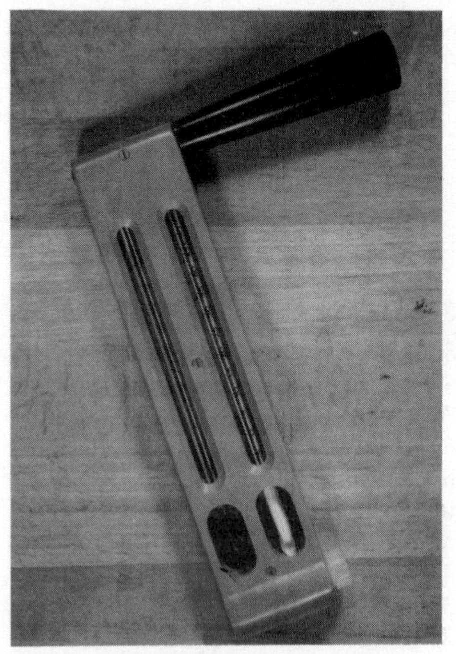

Fig. 53 Whirling hygrometer/ psychrometer. Photograph, courtesy of Brannan Thermometers Ltd

minute the rotation is stopped and the readings taken off the thermometers immediately. These are highly sensitive instruments, so much so that as soon as the air flows over it ceases the wet bulb temperature tends to rise as the reading is being taken and, since they are rarely fitted with radiation shields, the effect of sunlight may be significant. However, they are extremely useful, but need care.

Humidity can also be assessed by use of dial type relative humidity meters. These are of extreme value to the woodworker and being fairly inexpensive can be employed in various ways. Their use is discussed more fully in Chapter 10.

Air Circulation

Air circulating through a stack of wet wood serves two functions: one, it conveys heat to the wood, and two, it removes moisture from the wood in the form of vapour. The velocity of the air contributes to the rate at which the wood dries and may be (a) suitable to the species to be dried, or (b) too fast or unsuitable, in which case surface checking can result. Air circulation operates differently in air drying than it does in a kiln chamber, although the principle is the same.

In a kiln, the air passes over a heat source and is then directed by a fan or fans through the timber stack; this air is recirculated, but some of it, laden with moisture drawn from the wood is ventilated off, to be replaced by cooler air which now has to be reheated.

In the open air, if the circulation is sluggish the air will tend to become saturated and stagnant and drying will be minimal. During windy weather there will tend to be a continuous flow of air over the stacked wood, even so, there will be a film of air in direct contact with the broad surfaces of the wood which will move more slowly than the main flow and will have a higher vapour content.

When relatively dry air enters a pile of wood it picks up moisture and cools it as it passes from one side of the pile to the other; accordingly, the exit side of the pile will dry slower than the air entry side. In a kiln chamber, this is catered for by reverse speed fans, the air flow being reversed at predetermined periods, while a proportion is ventilated.

In air drying, not only does one side of a pile of wood tend to dry less readily than the other side, but the cool, moist air tends to fall so that the bottom of the pile is often less dry eventually than the top two-thirds. See Air Drying.

5
Kiln Drying

Various methods designed to improve the condition of wood have been suggested, experimented with, sometimes patented, and frequently discarded. One of the earliest recorded mentions of the endeavours of Man to season wood can be found in *Works and Days* written in 735 B.C. by Hesiod who was a Greek farmer and poet, a didactic work which was more of a moral essay on the dignity of labour than as a means of instruction, but he does record, "As the homes of the ancients were so smoky, it must easily be comprehended how, by means of smoke, they could dry and harden pieces of timber".

It would seem that smoke drying of wood was the forerunner of accelerated drying and indeed, the principle of using flue glasses to dry wood in kilns operated right up to recent years, the best known method being that of Bachrich. The Roman poet, Virgil, prompted, it is said, by Hesiod's *Works and Days*, produced in the years 37 to 30 B.C. *A Treatise on Agriculture* again a didactic work in four books. In Book 2 devoted to the cultivation of trees, Virgil comments thus: "Of beech the plough tail, and the bending yoke of softer linden hardened in the smoke". Attributed also to Virgil is the further reference: "These long suspend where smoke their strength explores, and seasons into use, and binds their pores".

These early methods were no doubt successful in rendering the 'seasoned' wood rather more stable. As timber dries, most strength categories increase and there is greater resistance to various forces. There is a good example to be seen in many of the Australian hardwoods where a preference exists among builders to use green wood for certain work simply because once the wood is dry it becomes difficult to nail. We have come a long way in our understanding of wood since Hesiod's day, but it has taken much experimentation and research throughout the world to arrive at our modern conclusions.

One of the timber drying kilns used in Europe and North America in the 1920's was the progressive kiln, so called because the wood, usually air dried first, progressed slowly from one end of the kiln to the other. The timber, stacked on wheeled bogies, entered the kiln broadside on, and as additional loads were placed in the kiln, usually at day intervals, they moved at this rate toward the exit and, below which were set a series of steam heated pipes. The bogies ran on a steel track set some five feet or so above the ground level of the kiln chamber. The kilns were provided with ambient air or 'natural draught' which entered the kiln through floor vents at the entry end and was drawn through the kiln, and the timber, towards the exit end by means of a tall chimney.

In effect, initial exposure of the wood was to cool, relatively dry air, but as each move was made, the air became warmer. The kilns held ten or twelve car loads but only about three began to come in contact with the warm air as they progressed to the final stage of drying. The heating pipes were controlled by valves set outside the exit end of the kiln as were facilities to inject live steam into the kiln if this were needed. The results from these kilns were generally good, but gradually they gave way to the more sophisticated compartment kiln which now relied upon forced draught supplied by fans and was a much advanced method of drying.

It is fair comment to say that the modern timber drying kiln involves a comparatively large initial investment, but long term, the economies effected will more than offset these charges. In recent years there has been an increasing interest in the use of dehumidifying units for drying timber and this aspect is dealt with later.

The advantages of employing a drying kiln are many; briefly, they include:

1. Since the drying elements are completely under control, any wood may be dried to a moisture content best suited to its use.
2. Kiln drying is more rapid than air drying; in kiln drying the seasoning time is reduced from months to days.
3. A drying kiln will provide a uniform and dependable supply of wood throughout the year regardless of weather conditions.
4. The temperatures used in most kilns tend to harden the resin in softwoods by evaporating the volatile matter, avoiding exudation in use.
5. Special sterilization treatments can be used to kill all fungal and insect infestations (including eggs and larvae as well as mature insects). This is particularly important in respect of *Lyctus* beetle which is a pest of timber yards, attacking the sapwood of hardwoods like oak.

6. All wooden containers destined for shipment to Australia must conform with Australian Quarantine Regulations which stipulates certain forms of treatment designed to rid the wood of any possibility of *Sirex* wood wasp attack. Recognised treatments include high temperature, high humidity sterilization. This can only be carried out in a timber drying kiln.
Commercial kilns fall into two major classes:

a) Conventional kilns operating mostly up to about 90 or 100°C using coal; oil; wood or other wastes as fuel.

The greater majority of the world's kilns are of this type, which are expensive to install and certainly not suitable for the craftsman with a small timber requirement. The typical kiln consists of a insulated chamber containing the necessary equipment:

i) Heating surfaces controlled at predetermined temperatures.
ii) Live steam injectors (o water atomizers) to control relative humidities at preset levels.
iii) Fans to circulate the heated and humidified air through the load.
iv) Ventilators, normally automatically controlled, to exhaust the evaporated water to atmosphere.
v) Measuring and governing and recorded devices by which the drying may be controlled.

b) Dehumidifiers.

There are a large number of dehumidifier kilns operating around the world, but generally handling only a very small proportion of the world's total timber volume. For example, in the United Kingdom there are more dehumidifiers than conventional kiln installations but dehumidifyer drying accounts for less than 5% of the total volume dried.

Dehumidifiers for timber drying fall into two main classes:

i) Low temperature units operating up to about 60° (with most operating up to about 50°C).
ii) Higher temperature units operating at 80°C and above.

The second type are suited to general commercial use, while the so called low temperature dehumidifiers are more suitable for the small business and the craftsman user to dry small parcels of timber.

The Conventional Kiln Chamber

Kiln chambers are sometimes brick built but more often today are carried out in aluminium modular construction and therefore portable in the sense of packaging and conveyance from the manufacturer to site.

The heating elements and fans for air circulation vary in their type and location depending on the design; some kilns simply accommodate a single load of timber, others may take two loads side by side. In some instances, the timber is loaded in packages by fork lift truck, in this case the loads entering broadside on.

Fans

Fans are made from non-corrosive aluminium. They vary in size, but may be as large as 72 inches (1.8m) in diameter, although 36 inch (0.9m) or 48 inch (1.2m) is more general. Control of air flow direction in older type kilns may be manually operated but usually is carried out automatically, the reversal of air varying in the time this is done according to the make of kiln. In some kilns, it is only necessary to make two changes each day, once in the morning, and again in the afternoon. In others, reversal of the circulation is necessary at shorter intervals, perhaps every two or three hours.

In kilns where large diameter fans are mounted on the kiln floor and directing air against load flanks, ie so called "side fan kiln", fans are reversed approximately every 30 minutes.

Air circulation may be affected by badly constructed loads. Care should be taken to ensure that adequate air flow exists through the whole timber stack.

Air Circulation and Velocity

Air circulation is required to remove the moisture picked up from the drying wood and to dispel it and to convey heat to the wood. The efficiency of the heating system of any kiln is increased by proper air circulation since, as the air velocity is increased, within the limits imposed by the wood in question, the heating elements give off more heat; in other words, air volume and its velocity is proportional to the drying rates of the species and thicknesses to be dried, and is at a minimum demand for a slow drying timber of low moisture content.

At one time, kiln engineers provided air speeds of 25 – 50 ft per minute for slow drying hardwoods and speeds in excess of 75 ft per minute were considered too high. Subsequent research has since proved that optimum speeds of 500 – 800 feet per minute (8 ft to 13 ft per second) are feasible, depending on length of travel, drying schedule, and the drying rate of green timber. Generally speaking, air velocities in kilns are nearer to 6 feet per second today since the higher speeds, while drying the wood faster, cost more in fan power. Another factor too, is where fairly thin stock is being dried, speeds say of 13 or 15 feet per second plus turbulence can disturb the top layers of wood in the stack unless these are weighted down. Where speed of drying is important and cost of power unimportant, it must be understood that the higher the initial moisture content of the wood, the greater will be the effect of higher air speeds. Wood that is partially dry, i.e. where the surface zones are near fibre saturation point on entry to the kiln, may suffer from too high an air speed.

Plenum Chambers

Air circulation not only carries heat to the wood and removes moisture from it in the form of vapour, it must also allow the vapour to escape from the vents and the air must be recirculated continuously. If the fans are situated to one side of the kiln, which is more usual today, then air will obviously enter one side of the stack and in a semi-saturated state emerge from the other. This means of course that one side of the stack will tend to dry more quickly than the opposite side; accordingly, the fans are arranged so as to permit reversal and allow the wood to dry more uniformly. Between the stacked timber and the walls of the kiln is a space or plenum chamber. As the air leaves the stack and enters the plenum chamber, it is deflected and directed back through the stack or stacks.

Provision of Heat

Conventional kilns may be heated by:

a) Steam fed from a central boiler fuelled by oil, gas, wood or other wastes. Generally, electricity, because of its high cost, is not used to heat conventional kilns.

b) Thermal fluid – a liquid fed to heating pipes at a high temperature, but at a low pressure. The low pressures used reduce many of the problems associated with high pressure steam. Thermal fluid, like steam, is heated by the same fuels at a central source and fed to individual kilns.
c) Oil or gas fired heat exchangers which are integral to each kiln. The provision of individual heaters to each kiln is more economic than a central boiler where only a few kilns are used.
d) Directly by the combustion of gas or oil. The flue gases together with the products of combustion are passed directly into the kiln chamber. The use of directly heated gas has been very successful in the U.K. and a number of commercial plants are in existence.

Kiln Recorder/Controllers

Instruments used to control the drying elements include dry and wet bulb hygrometers already referred to in Chapter 4. Ideally, these should be placed on each side of the kiln load so as to ensure the air is measured as it enters the stacked wood. With frequently reversed fans, controls on one side of the load are usual and adequate.

Modern kilns are normally fitted with automatic control instruments which not only control temperature and humidity to preset values but also operate the vents to ensure that no steam is wasted by the steam spray, used to regulate humidity, being operated while the vents are open.

Recorders are set outside the kiln and these record the wet and dry bulb temperatures within. A chart is placed in the recorder at the start of each kiln run and pens trace the air conditions prevailing. Thus, not only can the progress of the drying schedule be followed but the chart remains a permanent record of the run.

Drying Schedules

Research into the drying characteristics of different timber species has been going on for many years and, arising from tests in small pilot kilns followed by experimental full-size kiln runs 'safe' drying schedules have been evolved, that is to say, by following the official recommendations as to suitable air conditions and the appropriate times in which to make changes, the risk of serious degrade occurring in the wood is reduced to a minimum.

In some circumstances, say where a particular species is continuously being dried, a kiln owner will alter a recommended schedule in order to modify the drying but he will only do this after much cautious experimentation.

Kiln schedules are under constant review and new timbers are added as they are tested. Schedules are invariably supplied with new kiln installations, but they can either be supplied by the following organisations or advice and information given about them.

U.K.: The Timber Research and Development Association, Hughenden Valley, High Wycombe, Buckinghamshire HP14 4ND
U.S.A.: Forest Products Research Laboratory, Madison 5, Wisconsin.
Canada: Department of Forestry, Ottawa Laboratory, Ottawa.
Australia: CSIRO Division of Building Research, P.O. Box 56, Highett, 3190 Victoria.

It is important to recognise that differences will exist between a schedule produced in one country and that of another. Australia is a case in point. Many native hardwoods are refractory and schedules accordingly recommend special reconditioning treatment at the end of the kiln run: furthermore, in a good many cases, it is recommended that the wood is first air-dried to 30 per cent mc or less before kilning.

There is a tendency to use larger kilns in North America than in the U.K. and there is no doubt that American and Canadian schedules lean towards much faster drying times than we would employ. This is why, if modifications to any schedule is contemplated, it is unwise to adopt another authorities' recommendation without due regard to local conditions.

Kiln Run

When a kiln charge is ready, i.e. the timber has been stacked on the bogies and samples have been prepared and the initial moisture content of the wood has been determined, the bogies are placed in the kiln and the hygrometers placed in position. The fans are switched on and the timber and the kiln allowed to warm up. This can be done either by setting the dry and wet bulb thermometers so that a 9°F (5°C) difference is maintained until the desired dry bulb working

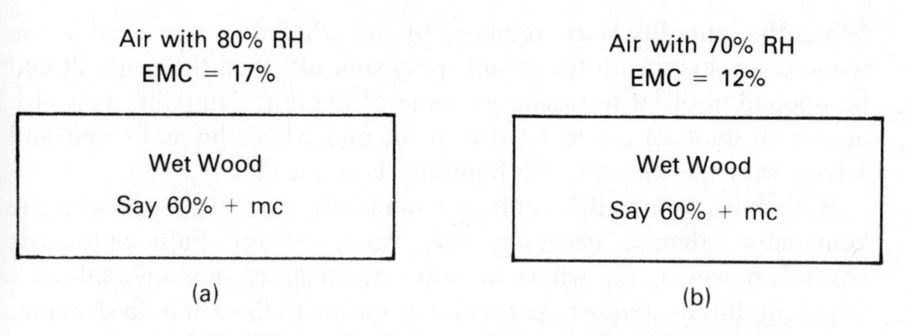

Fig. 54 If a 'safe' drying schedule demands 80% relative humidity, initially, reducing this may speed the drying rate, but could increase ultimate surface checking, since the equilibrium moisture content is now lower

temperature is reached, when the wet bulb reading is adjusted to that demanded by the schedule, or the controller may be set to the working dry and wet bulb temperatures straight away.

The important factor is the maintaining of the recommended relative humidity levels, especially in the initial stages of drying green timber; if the RH falls below the prescribed levels it can cause checking of the wood in the later stages of drying. Suppose the schedule calls for 80 per cent relative humidity at 38°C initially; reference to Table 7 shows this equals an equilibrium moisture content of about 17 per cent which is the level to which the timber would dry if left long enough in the conditions. If, however, the RH was allowed to fall to 70 per cent, the EMC would be about 12 per cent and obviously the wood would by drying that much quicker and probably initiating degrade problems which would only show up later. On the other hand, a few points variation in temperature would have only minor effect; if it were down the drying would slow slightly, if it were up, the reverse would apply, but so long as relative humidity is held at the proper level no problems should occur.

Venting

Provision of the right amount of relative humidity in the kiln is brought about by live steam injected into the chamber by spray jet.

When the humidity value required by the schedule is exceeded as the wood dries, even with the steam spray shut off, then the vents should be opened in order to discharge some of the hot, damp air. This also allows fresh, cool air to be drawn in, and when this is heated and mixed with the kiln air, the humidity is reduced.

With kilns where the vents are manually controlled opening the vents more than is necessary only wastes steam. Fully automatic control, however, i.e. where heating, steam spray and vents all react to set conditions, requires particular attention to the wet bulb sleeving, since if this is allowed to dry out then the vents will open and the steam spray will shut off, thus allowing the humidity to drop seriously. The water supply to the wet bulb and its sleeving must be regularly checked and remedied if required.

Drying Progress

The actual drying run is a routine linked to experience, since if the kiln is functioning properly then it is mainly the periodic checking of moisture content, changing charts, maintaining records, and making temperature and humidity changes in step with those in the schedule.

Much depends upon the actual kilning specification related to the timber species as to whether or not conditioning or case hardening relief treatments are applied towards the end of the kiln run. In some instances, notably with easy drying woods like obeche, lime, poplar and alder, conditioning is not resorted to, but in more exacting circumstances where close tolerances in moisture content spread are specified, or where stresses need to be ironed out, then equalizing and conditioning treatment is essential.

Conditioning establishes a more uniform moisture content. The procedures are as follows.

1. Dry the driest sample in the kiln to a moisture content 2 per cent below the desired final average moisture content. If this is say, 12 per cent, dry the driest sample to 10 per cent.
2. As soon as the driest sample reaches the moisture content stipulated, establish an equalizing equilibrium moisture content condition in the kiln equal to that value. This can be done by selecting a suitable combination of dry and wet bulb temperatures.
3. Allow the kiln to remain on equalization until the wettest sample reaches the desired final moisture content. If this is 12 per cent,

then the wettest sample should reach 12 per cent before stopping the equalization treatment.

Example: Assume a load of 1½ inch (38mm) thick beech is being dried to a final average moisture content of 12 per cent and towards the end of the kiln run the driest sample has a moisture content of 13 per cent and the wettest 16 per cent, and the kiln conditions at this stage represent 140°F (60°C) dry bulb and 114°F (46°C) wet bulb.

We allow the drying to continue until our sample with 13 per cent mc reaches 10 per cent and now we can equalize the load.

The kiln conditions at this stage represent an equilibrium mc for the wood of 6.5 per cent so to equalize the load we now need conditions that will give an equilibrium of 12 per cent, the desired final moisture content. By increasing the wet bulb temperature from 114°F (46°C) to 129°F (54.0°C) the humidity is increased to 73 per cent giving a suitable equilibrium value.

These air conditions will allow the driest and wettest portions of the load to draw closer, thus reducing the differences to a minimum. In a good many instances, at this stage it could be quite feasible to conclude the drying and allow the load to cool. Much depends upon the end use for the wood and the instructions covering the drying.

Case Hardening Relief

As part of the drying function, the wood may, as explained in chapter 7, become "case hardened". That is the surfaces are stretched larger than they should be.

If these surfaces were wetted any expansion would be resisted by the unwetted cores, stresses would be introduced producing compression sets in place of the tension sets caused in case hardening. If the surface compression set was equal to the previous tension set, then the timber would be expected to be free of stress.

In practice, case hardening stresses are removed by this very method. A high humidity, 90% or above, treatment is applied to the timber for a few hours at about the final kiln temperatures. Normally, a conditioning process would be applied to equalise the wood at the required final moisture content.

Often, such treatments are applied as routine procedures at many drying plants.

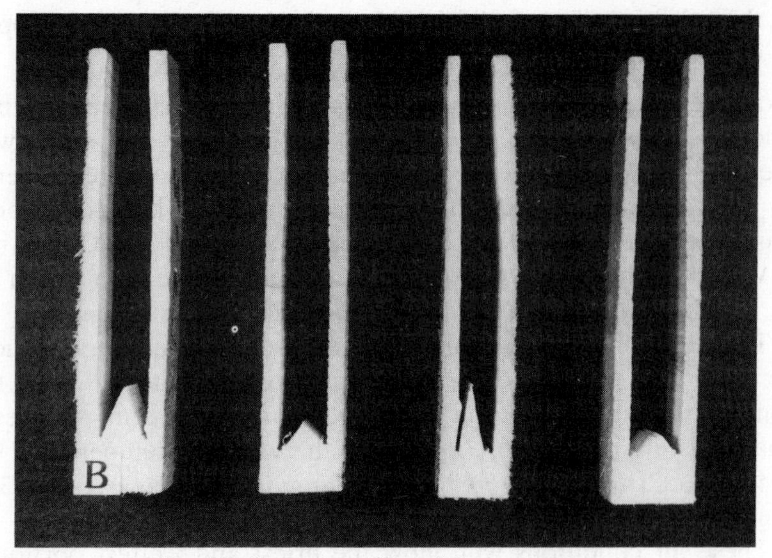

Fig. 55 Stresses relieved by steam conditioning. Photograph, courtesy of Peter Y.S. Chen

Cooling

When the kiln load is dry some thought must be given to cooling the wood. In some instances the load may be drawn from the kiln almost immediately but it must be recognised that at the end of the run the wood will be hot and if removed to the outside air this will circulate through the load and become dry. Under these conditions surface checking could occur, especially where the working temperatures had been high.

If the heating and fans are switched off the load can be left to cool without undue difficulty.

Drying Wood on a Small Scale

The small craft workshop and the home worker often have the greatest difficulty in obtaining properly dried wood. Sadly, in many cases, the difficulties are such that often the wood is used inadequately dried.

Frequently, it is said that, historically, fine joinery and furniture was made long before kilns were invented, and in those days the wood was properly seasoned before use. In practice, the long periods of air drying, created by the extensive durations of "seasoning" were followed by storage in unheated buildings and workshops. Timber was produced suitably dried for the intermittently partly heated environments of the times. The inadequacy of the historic seasoning is very apparent when viewing old joinery and furniture exposed to modern centrally heated environments.

The long periods of initial seasoning had no effect whatsoever on the ultimate stability of the wood when in service.

Nevertheless, a similar, but more effective method may be followed today. That is, air drying may be followed by periods of storage, firstly inside unheated buildings and subsequently stored in rooms which are heated.

Any system which relies upon air drying must include the following disadvantages.

i) No, or very little, control of heat, humidity and air circulation.
ii) Drying is governed by ambient conditions, sometimes final moisture contents may be about 15% but are more likely to be about 18–20%.
iii) The durations of air drying may be delayed and unpredictable.
iv) Because of the difficulty in controlling drying, it may be impossible to avoid degrade. Insect, and fungal attacks such as blue stain, will continue in the stack until the wood moisture content falls to about 20% or so.

To advance beyond simple air drying obviously requires a greater sophistication of control and of course, inevitably increased expense.

To be effective, a kiln, however small, would need:

i) A chamber – properly constructed, protected from the weather, fully insulated and unaffected by warm vapour.
ii) Air circulation – possibly by natural convection, but much more preferably by means of a fan or fans.
iii) A source of heat – which must be evenly distributed throughout the kiln and capable of being controlled at all temperature levels.
iv) Extraction of evaporated water – the removal of vapour from the kiln atmosphere may be by ventilation or by dehumidification.
v) The means of controlling relative humidity.

It must be realised that the small scale woodworker, who wishes to dry his wood must compromise between:

i) Initial cost.
ii) Speed of drying.
iii) Effectiveness of drying.
iv) Running costs.

Generally, the faster, higher quality and more economic the drying the higher the initial cost of the equipment and the lower the running costs per unit of moisture removed.

Generally, the lower the sophistication of the equipment, the more drying durations become extended and running costs to rise. The main quality, in terms of degrade and moisture distribution, tends to fall. From this, it will be seen that the craftsman will inevitably pay more for his drying than a large timber producing and manufacturing company.

Considerations in the choice of a small kiln.

Probably the greatest difficulty facing the builder of a small kiln, is the provision and, more importantly, the control of the necessary heat and humidity.

The small low temperature dehumidifier operating up to about 50–60°C probably offers the best compromise in equipment for the craftsman user. However, the equipment must be properly installed in an adequately constructed chamber. The water extraction rate of the dehumidifier must be adequate for the volume of water that is to be removed from the wood in the required time.

Generally, softwoods dry more rapidly, and contain more water, than hardwoods. Thus, more water has to be removed in a given period with softwoods than with hardwoods. For this reason, dehumidification requirements should be calculated on the basis of softwood requirements where softwood or softwood and hardwoods are to be dried. For example, 50mm pine is likely to require over four times the water extraction capacity than when drying say 50mm English oak.

Whenever possible, before embarking on a small scale home built kiln, it is preferable to contact suppliers of dehumidifiers specialising in timber drying and also, if possible, to obtain independent advice (See Fig. 56).

The use of small dehumidifiers overcomes, or reduces, the difficulties associated with the provision and control of heat, humidification and ventilation, all of which otherwise require costly equipment. Drying of the thicker, more permeable woods, may be extended in duration and difficult, when using low temperature dehumidifiers.

Occasionally, in a craft or woodworking magazine a kiln design is offered as suitable for home or small workshop use. These should be treated with extreme caution and advice sought before proceeding. In the U.K. advice may be obtained from Question Columns in magazines from the Timber Research and Development Association and often from timber companies who have their own kilns. Some of these, contrary to popular belief will often be helpful to the craftsman, so suitable approaches are always worth while.

Very small, home built, conventional kilns, other than expensive small scale laboratory equipment, seldom operate satisfactorily. Nevertheless, for those wishing to embark on such a project a simple drying kiln may consist of:

i) A low temperature heat source, e.g. hot water pipes with loads suitably stacked to make use of convection currents.
ii) Ventilators at floor and ceiling level, which are open or closed when the relative humidity varies from the desired levels. Fans may be added to create a more effective air circulation.

   At low temperatures, say about 20°C, a relative humidity of about 80% could be maintained, until the wood moisture content fell to about 18%. At that point a relative humidity of approximately 55% would eventually reduce wood moisture contents to around 12%. If signs of unwanted checking are perceived then the relative humidity levels should be raised.

Drying would be likely to be slow and require very frequent checking. Also, due to the low temperatures used, difficulty would be experienced in drying the thicker and less permeable woods.

Another approach, would be to open stack properly air dried wood in a centrally heated room where temperatures could be maintained at approximately 6 to 7° above ambient.

The wood should remain in the room until representative boards, when periodically weighed, show no further loss in weight. If any surface checking is noticed the room temperatures should be reduced, and, if necessary, timber covered for a period to inhibit air flow.

It scarcely needs mentioning that kilning should be conducted on sawn lumber and prior to any reduction by hand or machine cutting.

In this way any slight surface imperfection of superficial staining or checking etc may be removed.

Drying-chamber Construction

Building a chamber for a small-scale wood dryer is relatively straightforward. It is likely to be one of the easiest projects encountered by a cabinetmaker. Wood-dryer manufacturers offer free advice to users on overall dimensions and layout so that you can be assured of a system built to your requirements.

A small drying cabinet which can hold 35 cu ft of wood may measure 8ft (2.44m) in length, and 4ft (1.22m) in height and width. The walls, floor and top of the chamber should each consist of a softwood frame clad on both sides with exterior-grade plywood. The cavity should be filled with expanded-foam board for insulation. The dryer manufacturer will advise on suitable insulation thicknesses in individual cases, since these depend on the surface area of the chamber. A chamber of the dimensions given may incorporate 2ins (50mm) of insulation.

The walls must also incorporate a vapour seal to prevent saturation of the insulation and the ingress of moisture to the chamber. Vapour-proof paint finishes are available which can be applied to the interior of the chamber; alternatively, heavy-gauge polythene sheet should be attached over the insulation before the the inner cladding is fixed in position. Taken care to ensure that all joints are taped and sealed and the exposed nail heads are covered. A close-fitting removable door panel should be incorporated in the side of the chamber, through which the timber will be loaded. Fit closed-cell synthetic-foam strip around the door opening, together with retaining catches.

De-humidification wood dryers small enough to be used by the cabinetmaker and craftsman joiner are readily available. Several thousand are currently in use today. The smallest systems will dry upwards of 20 cu ft per load.

Modern Developments

Kilns are now available where the functions of temperature, relative humidity, air circulating, venting, wood moisture content, schedule variation etc, may be controlled remotely by micro processor units.

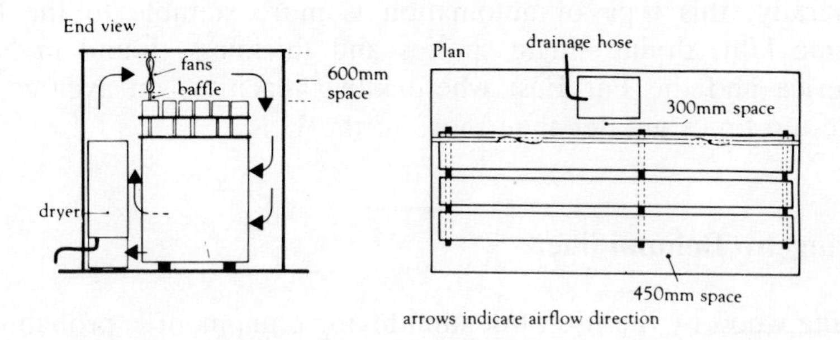

Drying-chamber layout for large machine

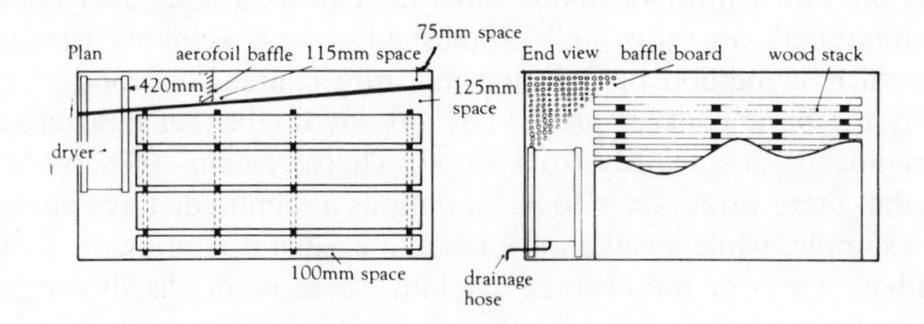

Drying-chamber layout for small machine

Corner of chamber (plan section)

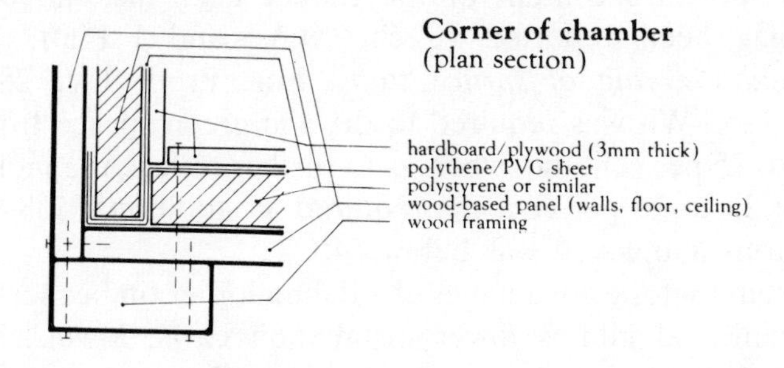

Sealing door frame

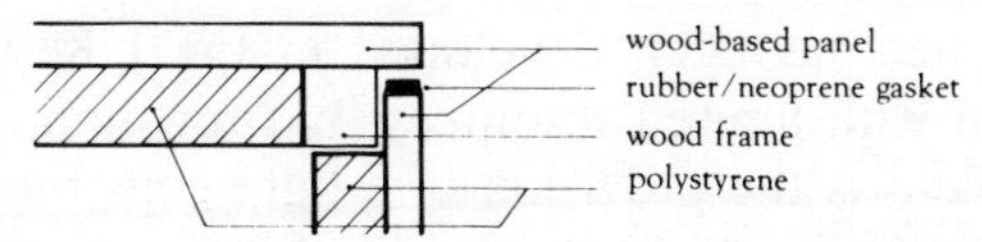

Fig. 56 Drying chamber construction
By kind permission of Ebac Ltd

Generally, this type of automation is more suitable to the large volume kiln, drying single species and thickness, found in North America and the Far East where kiln capacities, on average, are some ten times greater than those in the U.K.

Drying by Dehumidifier

Drying wood by the use of dehumidifying equipment is probably the most satisfactory alternative to orthodox kiln drying and more particularly at the present time bearing in mind the high cost of fossil fuels on which most orthodox kilns depend as a source of energy. Dehumidifiers are now well established as a reasonably economic and efficient method of reducing moisture content in wood; that is to say the basic concept proves beyond any doubt that moisture can be removed satisfactorily from wood. There remains, however, the fact that there would seem to be limitations not immediately apparent. For example, while installation costs and ease of operation are factors of advantage over the average dry kiln, there is the disadvantage of slower drying and greater energy consumption when timber is dried below 15 per cent mc because of the low drying temperature used and the low moisture content of the timber near the end of the drying. It has been reported (Cech; M.Y. and F.Pfaff; 1978; *Dehumidification drying of spruce studs*. For. Prod. J.V. 28 (3): 22–26.) that 125 kWh was required to dry the green spruce from 52 per cent mc to 25 per cent mc, whereas to further reduce the moisture content from 25 to 16 per cent mc required an additional 138 kWh. The temperature employed was below 54°C.

It would seem that the advantages of a dehumidifier timber seasoner over a conventional kiln is lower installation costs, lower energy requirements when drying from green to about 20 per cent mc, less drying degrade, and easier operation, while the disadvantages are slower drying and greater energy consumption in drying timber to below 15 per cent mc.

A small Mini-Seasoner has been developed in the UK which is ideal for the craftsman with limited resources, but for more general use the basic requirement is the provision of a tight constructed and well-insulated chamber into which the timber and one or more dehumidifier units can be accommodated. Such a chamber can be erected within a small workshop for convenience and, unlike a conventional kiln, where moisture in the form of vapour is vented

Fig. 57 A commercial single stack dehumidifier kiln installation by Cubbage-Bollmann Ltd

out of the chamber into the surrounding air, in a dehumidifier, liquid water is drained off as condensate.

As previously explained, heat is necessary to evaporate water. Required heat is in two forms. More heat is required to cause evaporation than to raise to boiling point. For example, on one kilogram of water raised from 0°C to 11°C would require approximately 419 kilojoules. To evaporate the kilo of boiling water would require a further 2257 kilojoules. The extra heat required for evaporation is known as the latent heat.

When timber is dried in a conventional kiln the evaporated moisture is exhausted, in vapour form, through the kiln ventilators and replaced with cooler, dryer air from outside the kiln. The latent heat used to evaporate the moisture is lost, and extra heat is required to warm the incoming air to the kiln temperature. In drying by means of dehumidifiers a thermally more efficient method is employed to remove moisture. Drying is undertaken in a sealed chamber and there is no deliberate interchange of internal and external air.

The moisture evaporated from the wood is condensed on the evaporator coil of the dehumidifier and is drained away as water. As the vapour condenses latent heat is passed to the dehumdifier refrigerant.

This latent heat is recovered as the recirculating air passes over the condensing coil of the unit. The recovered latent heat is then used to evaporate more moisture from the wood.

Fundamentally, with a conventional kiln a quantity of heat is used to evaporate a volume of water. That quantity of heat is then discarded with the volume of evaporated water. Alternatively, in a dehumidifier the same quantity of heat is used to evaporate a volume of water and then recovered for evaporating further volumes of water. Obviously, if dehumidifiers are installed and used properly the heat required for a given amount of drying is considerably less than by conventional methods.

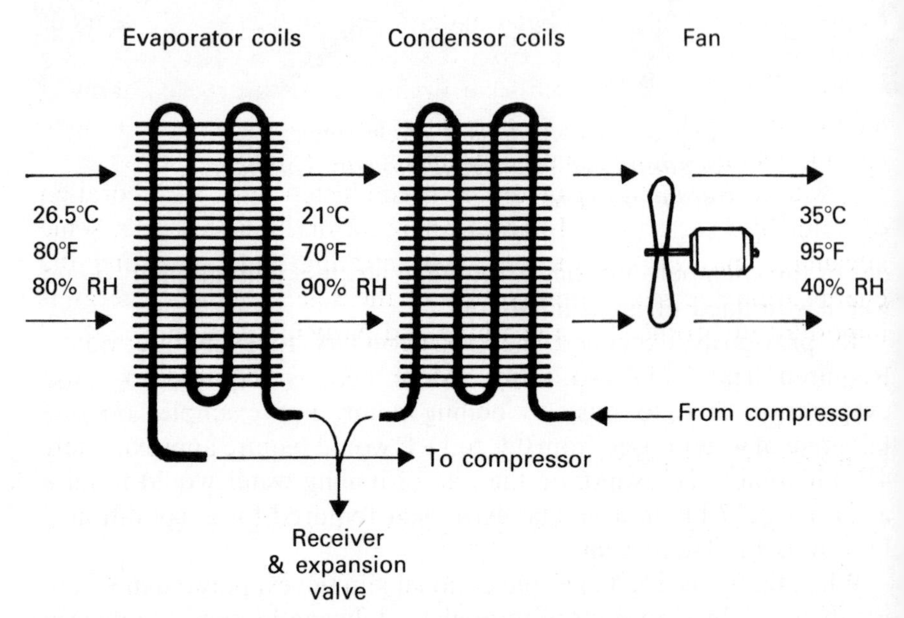

Fig. 58 Diagram showing air condition changes as it passes through a dehumidifier

It must be appreciated that the principles of drying by dehumidification are the same as those in the orthodox method of air and kiln drying. The difference is in the disposal of vapour after the water has evaporated from the wood and the differences in thermal efficiency.

Much depends upon the demands made on a dehumidifier dryer as to its economy of drying. If green hardwoods in relatively thick dimensions are continuously to be dried, then the dryer could be tied up for lengthy periods because of the low temperatures employed; but on the other hand, drying degrade would tend to be less than might be the case with conventional drying. Drying partially air-dried timber down to a final mc of 10 per cent is probably the most economic performance although lower final moisture contents can be achieved. An advantage to the small consumer of wood is that mixed parcels can be dried, but a distinct advantage is the quite small dehumidifying units presently available for a relatively small cost.

Solar Drying

Considerable research has been undertaken over the last 25 years or so on the direct use of heat from the sun for drying wood.

Obviously, any value from solar drying will be greater in areas of high sunshine. The potential availability of solar energy depends upon latitude, cloud, and the time of year.

It will be realised that air drying is conducted by the evaporation of water by solar energy. In areas where air drying is possible, some advantage may be obtained by concentrating and conserving the energy from the sun. Conservation of the solar obtained energy is important in periods of no sunshine and particularly at night.

General Developments

In spite of many years of research and the construction of numbers of proto-type and experimental plants, there are very few solar kilns operating commercially. Of course, this does not mean that the small scale user could not find some advantage by cautious experimentation.

There are two main types of solar kiln:

a) The so called green house type. The structure is similar to a green house but the equipment and operation falling into various categories.

   i) Simple enclosure with circulation and venting by convection as within a traditional green house.
   ii) Constructed as in i) but the air circulation is provided by fan.
   iii) Fans together with humidity and ventilation control provided.
   iv) As above but with internal integral solar energy collectors.

b) Kilns with external solar collectors, either fully heating, or supplementing conventional heating sources. Obviously, greater sophistication in forms of control, circulation and ventilation may be used. As an alternative to ventilation, vapour may be removed and the latent heat returned to the chamber by dehumidification.

The green house type of kiln is limited in capacity because the area for solar collection decreases as the volume of wood becomes greater. Increased timber volumes require external solar collectors, while the need for good quality drying and the reduction of degrade requires means of controlling the heat, humidity, circulation and ventilation. The increased expenditure, to obtain such control, may raise costs above those of conventional kilns. If the craftsman or small user is interested in solar drying there is substantial literature on the subject, but it must be realised that most of the work is experimental. All relevant work to about 1981 is summarised in "Solar Kilns for Timber Drying the State of the Art" by Timber Research and Development Association, High Wycombe, Bucks, United Kingdom. A comprehensive list of references is included which is obtainable from library sources.

A few research workers have offered designs which they claim are suitable for the small business.

For example: a small 1 m^3 (35.3 ft^3) solar kiln designed and developed at Oxford and intended for the small user is shown in Fig. 59.

American Developments

Various studies have been conducted in America on solar drying and of particular interest to the small business is a solar heated timber

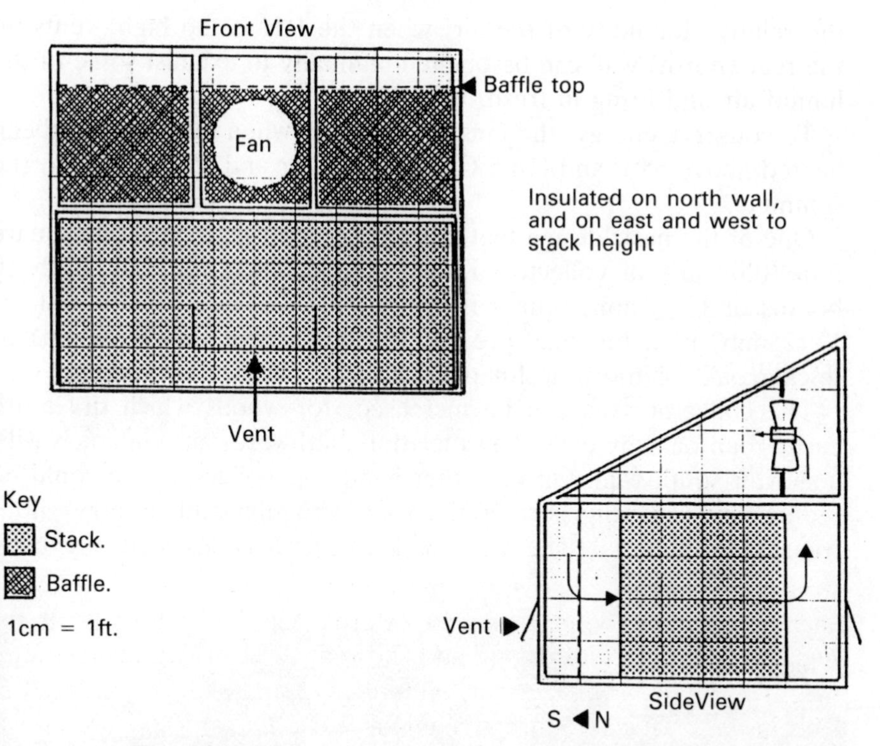

Fig. 59 1 cubic metre capacity kiln for D.I.Y. and small carpenters

dryer designed, constructed and tested at Virginia Polytechnic Institute, and State University by Eugene M. Wengert, Associate Professor, Wood Technology. This dryer can hold up to 1500 board feet (235 ft^3 or 3.5 m^3) of 1″ (25mm) thick timber per charge, and has been designed to dry a charge in approximately one month of moderately sunny weather in the mid-latitudes of the United States. Presently, there are over 250 units based on this design operating in the USA, Brazil, Jamaica, and elsewhere.

The dryer incorporates a passive collector, similar to a greenhouse. There are four insulated walls and an insulated floor. The roof is clear plastic or fibreglass (two layers work better than one) sloped at a 45° angle to the south.

Solar energy enters the dryer through the clear roof and is incident on one of the black-painted interior surfaces. The energy is converted to heat and this heat is, in turn, circulated through the timber pile where it is used to evaporate water. The evaporated water increases the relative humidity of the air; when the RH is too high, vents on the rear (north) wall can be opened manually to exhaust some of the humid air and bring in fresh, dry air.

To conserve energy, the fans are run only when the dryer has been heated above 65°F and (18.5°C) in the winter and 75°F (24°C) in the summer.

One of the basic design features of this dryer is to have one square foot (0.09 m^2) of collector, i.e. of sloped, clear roof, for each 10 boards of 1″ (25mm) timber in the dryer. This ratio works well for 1″ (25mm) oak, but may provide too much heat for 2″ (50mm) or thicker oak, or too little for pine or other fast drying wood.

The collector area can be increased, for woods which dry more easily than oak, by extending the roof southward, accompanied with a shorter south wall. On the other hand, the collector area could be reduced by covering part of the roof with plywood or other non-transparent material. The Virginia Tech Dryer is designed to control

Fig. 60 The Virginia Polytechnic Institute and State University solar dryer designed for small business. Photograph, courtesy of Eugene Wengert

expenses and yet provide a functional, reliable design which will dry a charge of timber every month under American conditions.

Construction Details

The floor is framed with 2″ × 6″ timbers set at 16″ centres (50mm × 150mm at 406mm) in order to be strong enough to carry the weight of green timber. If softwoods are used for this purpose they should be preservative treated if in contact with the soil. If a hardwood such as keruing/gurjun is used then a couple of brush coats of an organic solvent wood preservative should be applied to the wood and the depth of the bearers could be reduced to 5″ (127mm).

All the framing members are of nominal 2″ x 4″ (50mm x 100mm) softwood. Paper backed fibreglass insulation is installed between the framing members. The interior and exterior are covered with ⅜″ (9mm) exterior grade plywood. The inside plywood is painted with two coats of aluminium paint, which, when dry, acts as an excellent vapour barrier. A third coat of flat-black paint is applied to the interior.

The exterior is also painted, but a vapour barrier should not be used; if any moisture does migrate into the walls, it must have an easy way to escape to the outside. For this reason, paper-backed insulation is considered better than foil-backed insulation.

The roof is framed on fairly wide spacing. This spacing may have to be adjusted to accommodate the width of the covering material and any anticipated snow loads. The clear covering can be transparent polythene or other plastic sheets, but these coverings do deteriorate fairly quickly with heat, ultraviolet light, and continuous fluttering and flexing due to air turbulence from the fans. The most durable covering would be ordinary, almost transparent, corrugated fibreglass, available from most builders' merchants. Two layers of covering, separated by a dead-air space, are suggested.

The dryer has two access doors at each end (east and west walls) to permit periodic examination of the timber and measurement of moisture content. In addition, the Virgina Tech Dryer has the roof hinged to the north wall and the south wall hinged to the floor. This permits the roof to be raised and south wall lowered to facilitate loading and unloading.

The vents on the north wall are framed openings with a small piece of plywood that acts as a door.

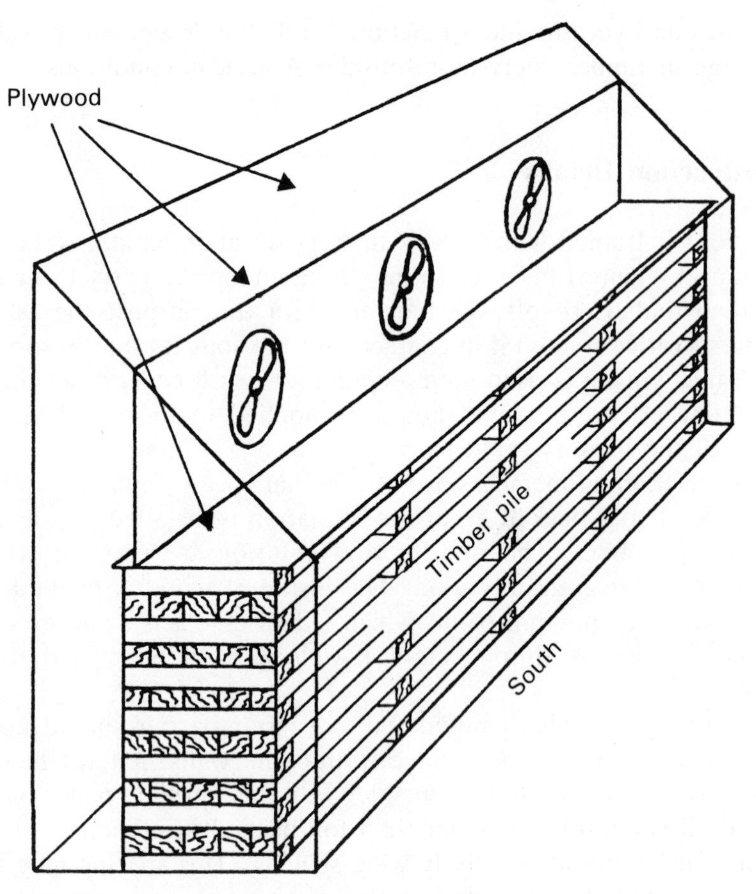

Fig. 61 A sketch of the solar dryer shows its simplicity. The roof is clear plastic or fibreglass; the walls are framed with 2″ × 4″ (50 × 100mm) stock with insulation and plywood covering; and the floor is 2″ × 6″ (50 × 150mm) stock with insulation and plywood covering. The interior plywood walls, baffles and fan housings are painted black to absorb maximum solar energy.

The fans are 3-speed window fans with thermostatic off-on control. They are fastened to the roof framing about 18″ (450mm) in front of the north wall with a plywood shroud of baffle around them extending downward 3 feet (1m) below the roof and running the full length of the dryer in order to force the air through the timber pile. Whenever the dryer is left empty and fans not operating, the roof should be

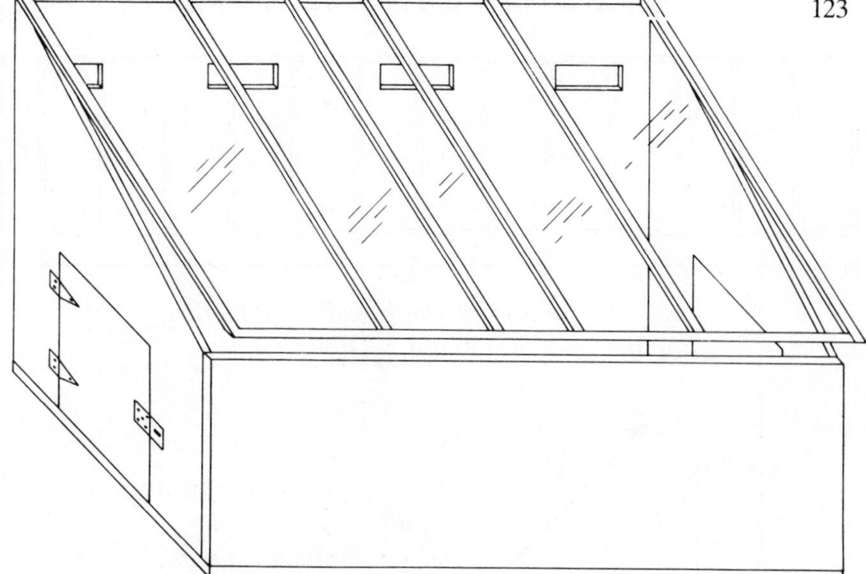

Fig. 62 Sketch of dryer assembly, with roof opened slightly, shows overall assembly. The roof is hinged at the top to the north wall and the south wall is hinged at the bottom to the floor so that the dryer can be opened for loading. Doors in the east and west walls permit daily inspection.

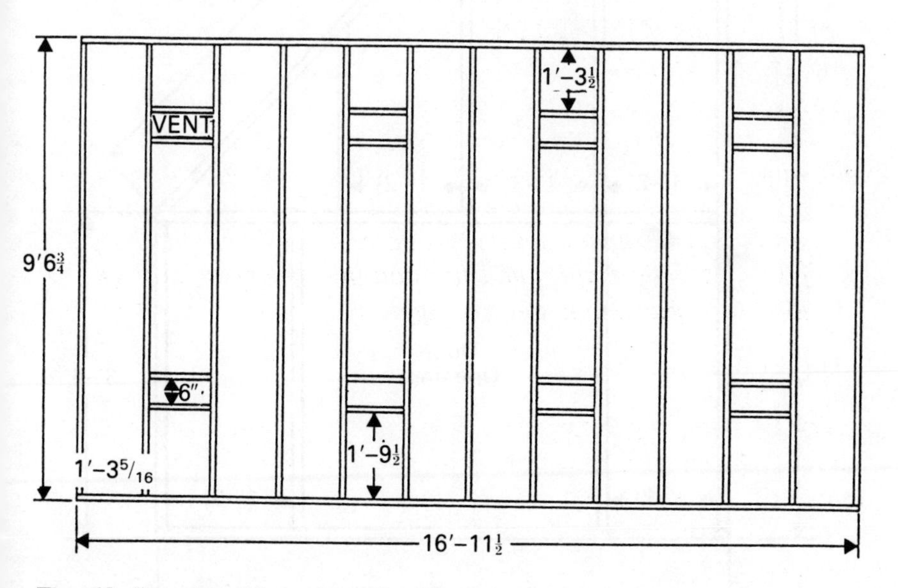

Fig. 63 Framing of north wall. All timber is nominal 2″ × 4″.

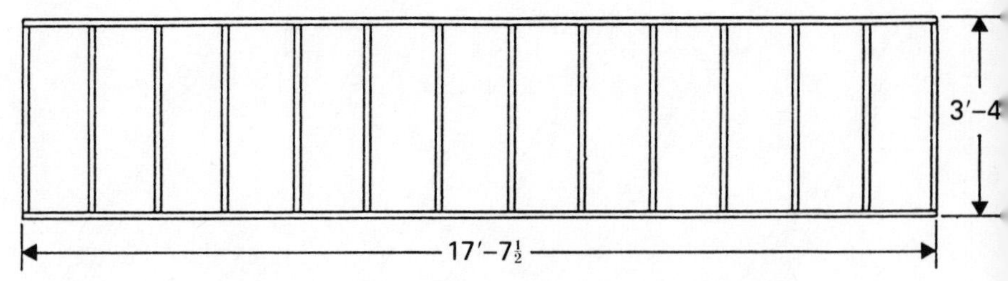

Fig. 64 Framing of south wall. All Timber is nominal 2″ × 4″ (50mm × 10mm)

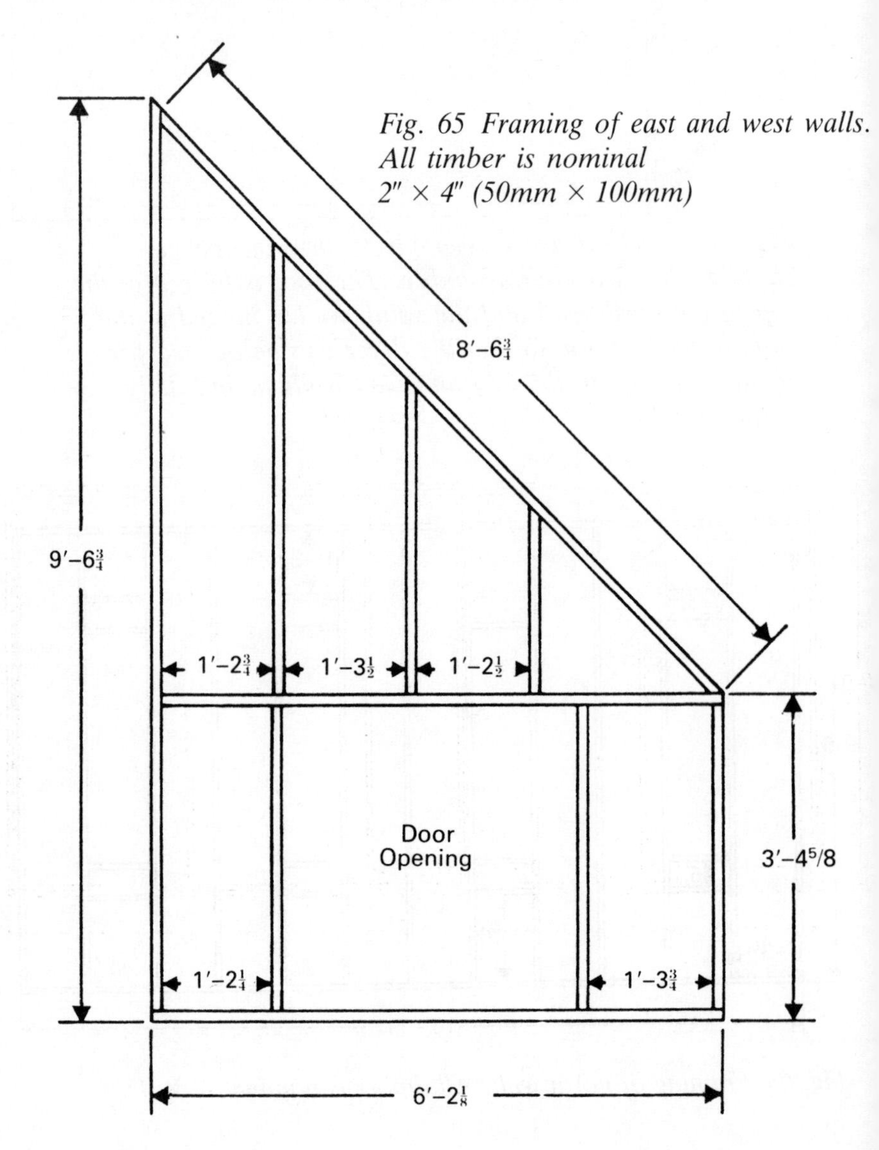

Fig. 65 Framing of east and west walls. All timber is nominal 2″ × 4″ (50mm × 100mm)

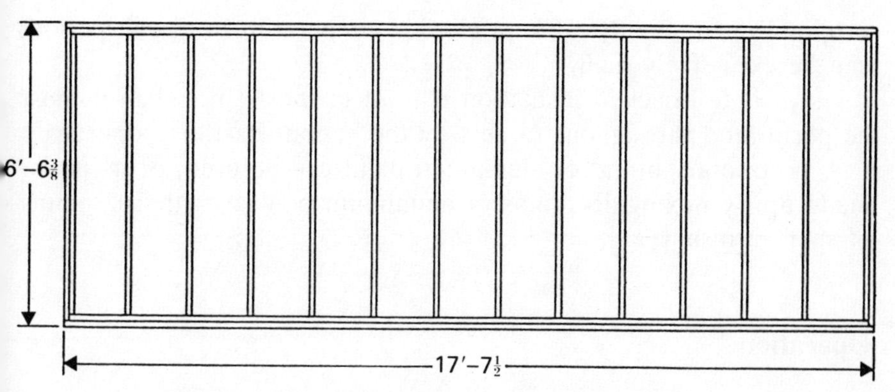

Fig. 66 Framing of floor. All timber is nominal 2″ × 6″ (50mm × 150mm) (This may have to be increased to 2″ × 8″ (50mm × 200mm), for heavier loads or long spans.

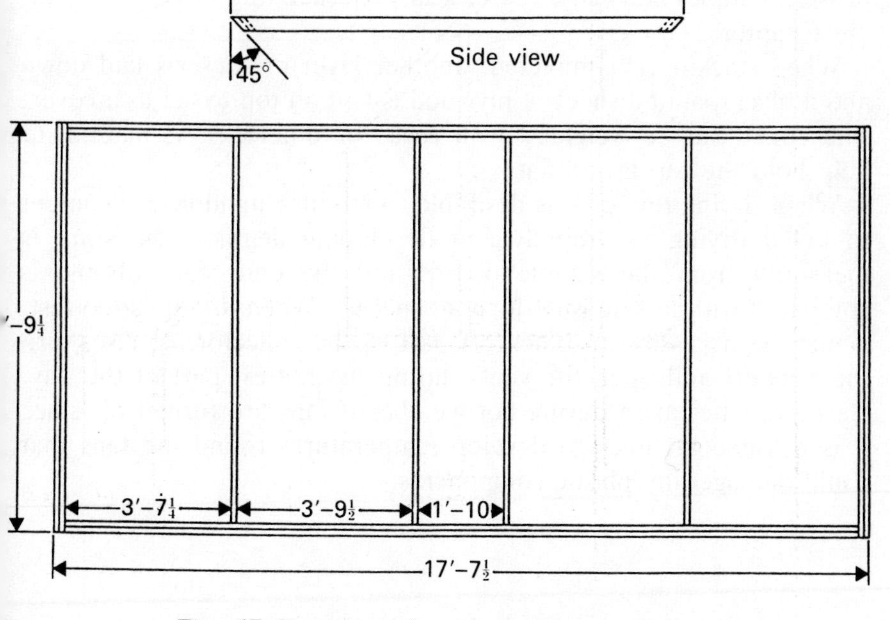

Fig. 67 Framing of roof. Spacing may be altered to accommodate width of covering material or heavy snow loads. All timber is nominal 2″ × 4″ (50mm × 100mm)

propped open slightly. If kept closed, temperatures at peak periods can be excessively high.

Note: If foil-backed insulation is used in the frame, then it must be perforated throughout to destroy the vapour barrier properties.

A proprietary brand of aluminium paint can be used, or it can be made up by mixing 18 grams of aluminium powder with 102 grams of spar varnish.

Operation

Green timber to be dried should be end coated with aluminium paint or other commercial end sealing compound immediately after sawing to prevent end checks and splits from developing. End coating is not too effective when applied after the timber has begun to dry. The timber should then be stacked in the kiln with one foot (300mm) clearance on either side of the stack to permit adequate air circulation, and in neat layers separated by sticks, as described in Chapter 6.

While the timber is being stacked, several boards must be cut to provide samples that can be periodically checked for moisture content; see Chapter 2.

When stacking is completed, another layer of sticks is laid down and a black-painted sheet of plywood is laid on top to act as a cover. The cover can be weighted with rocks or other heavy material to help hold the top layers flat.

When drying timber it is desirable to monitor its moisture content to avoid drying too rapidly and developing degrade. Moisture is measured from the sample boards and the daily rate of loss is compared with a 'safe-rate' for that species. When drying is too fast, it may be necessary to block off part of the collector, or else, turn the fans off and open the vents during the hottest part of the day. Care must be taken during hot weather if fans are turned off since it is possible at times to develop temperatures round the fans that could damage any plastic components.

6
Air Drying

Seasoning wood in the open air is simply a means by which as much as possible of its excess moisture can be removed. This somewhat hypothetical amount, deemed to satisfy the requirements of end use is a variable factor according to how much moisture a finished wooden item can tolerate.

In temperate climates, air dried timber with a relatively uniform spread of moisture throughout its bulk is usually suitable for all external uses and for interior use in areas such as roof voids and floor joists. It does not follow, however, that the flooring ultimately to be laid will be satisfactory if produced from air dried stock without some further drying. There are valid reasons for this: structural timbers, when placed in new buildings, are initially subjected to fairly moist conditions, even with timber framed buildings, due to various forms of moisture introduced into the building processes, and it means that a degree of movement in these members has to be generally accepted; there is, however, a certain amount of restraint to movement offered by the general layout and construction of joists, purlins or trusses.

In the case of flooring, softwood t & g boards are generally laid quite early in the building process, but since much of this, at least in the UK, had some form of forced drying, often by kilning, prior to shipment, at the time of laying its moisture content is relatively low, so movement is often fairly moderate. In the case of hardwood flooring, usually of parquet strip or mosaic squares, these are invariably produced from kiln dried wood and no experienced flooring contractor will lay these until the sub-floor has dried out thoroughly.

There is the point too, that movement by shrinkage in structural timbers is usually greater in a vertical direction because most sections are placed on edge, whereas movement across a floor takes place horizontally.

Interior structural timber could be called non-exacting, so that good air dried timber is satisfactory, but the same does not apply to

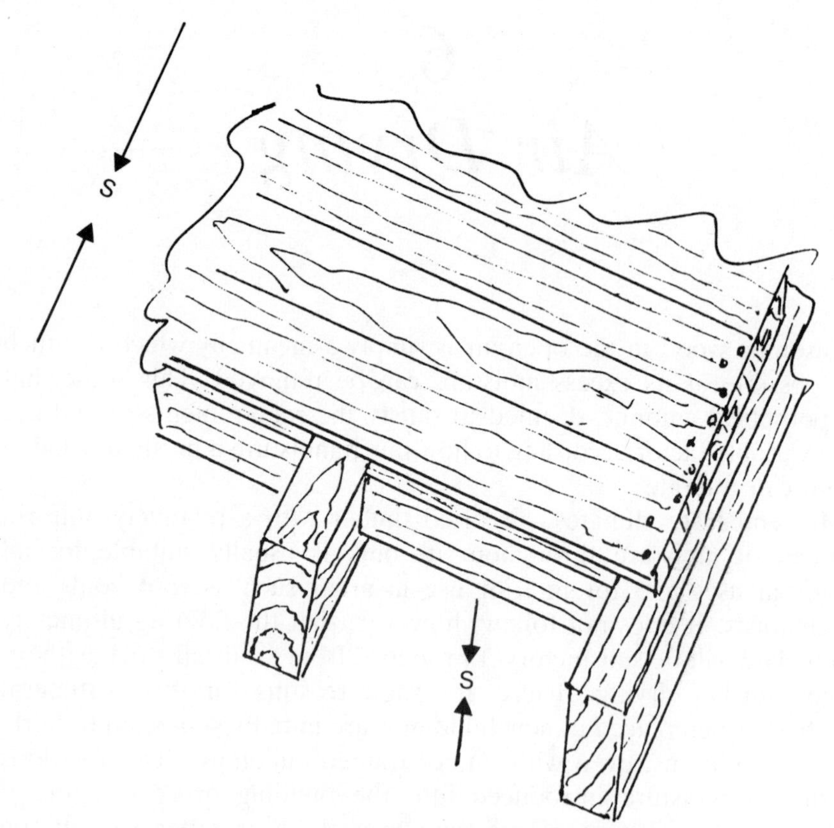

Fig. 68 In constructions such as suspended floors, potential shrinkage (S) in joists & noggins is opposed to that of the floor, thus helping stability of the structure.

other interior woodwork such as furniture, cabinet making, high-class joinery and so on; these uses are exacting and they apply not only to domestic premises but to all well-heated buildings, and in those circumstances excessive movement by shrinking and swelling of wood should not be tolerated; accordingly, air dried wood must undergo further drying before its condition can be regarded as satisfactory.

Generally speaking, no matter how long timber has been stored in the open air, seldom, if ever, will it be dry enough for immediate use for exacting interior uses. That is not to say it cannot be brought in and some work commenced, but recognition must be paid to just how much moisture the wood holds at that stage and how much more must be removed, and by what means, before any thought can be given to the direct manufacture and assembly of the goods. Wood extracted from a yard during the summer will be at least 5 points of moisture higher than the indoors equilibrium, and in winter probably twice that amount. Air drying is an important step in the preparation of wood and it should be studied and applied carefully. Table 26 indicates suitable moisture contents at time of installations. Fig. 69 gives equilibrium values ultimately.

Climatological Conditions

Wood placed in the open air to season will be subjected to variations in temperature and relative humidity not only according to seasonal changes, but also during each twenty four hours, and since the atmospheric conditions at any given time correspond to an equilibrium moisture content for wood, the rate at which the wood will adjust will depend upon how long those conditions are sustained. In effect, by the time the wood has warmed after a cool night, the actual drying period each day may be condensed into just a few hours. The following tables give a summary of climatic conditions and their effect, in different locations.

It will be seen that the average relative humidity throughout each twenty four hours is in excess of 80 per cent for a large part of the year, equalling a drying equilibrium for wood of more than 17 per cent, while the best drying periods are during the daylight hours from March until September, but taking an average for the whole year, relative humidity is 79.5 per cent and equilibrium moisture content 16.4 per cent, which in effect is about the driest one can expect airdried wood to achieve in this type of situation. Air conditions and corresponding EMC for other locations are summarised below.

The annual averages given are based on Meteorological Office data covering a ten year period, i.e. 1961–1970 and it will be seen that taking the British Isles as a whole, equilibrium moisture content values for wood stored in the open air are a lot higher than is generally realised. The summer months do provide lower equilibrium values and obviously have a bearing on the condition of wood, deemed to be air dry, if extracted from the yard during this time. The following table gives a guide to summer conditions.

It should also be noted that equilibrium moisture content is directly related to relative humidity, while temperature governs the moisture

TABLE 10:

Locality: London

Time (GMT)		Jan	Feb	Mar	Apr	May	Jun	July	Aug.	Sept	Oct	Nov	Dec	Average
03	Temp °C	3.1	3.1	3.8	6.1	8.9	11.9	13.5	13.4	12.0	9.8	6.0	3.4	7.9
	Rel. Hum. %	90	87	87	89	88	87	87	89	91	91	89	89	89
	EMC %	20	18	18	19	19	20	20	20	20	20	19	19	19
09	Temp °C	3.1	3.4	5.1	8.6	12.4	15.9	16.7	16.2	14.2	11.1	6.4	3.5	9.7
	Rel. Hum. %	89	86	81	79	72	70	72	78	85	87	88	89	81
	EMC %	19	18	17	16	14	13	14	16	17	19	19	20	16.8
15	Temp °C	5.3	6.2	9.1	12.0	15.6	19.5	20.3	20.0	18.2	14.7	9.0	5.4	12.9
	Rel. Hum. %	81	73	65	63	59	56	58	60	64	70	77	82	67
	EMC %	17	14	12	12	11	10	11	11	12	13	15	17	13
21	Temp ° C	3.7	3.9	5.6	8.5	12.0	15.9	16.8	16.2	14.2	11.3	6.7	3.8	9.9
	Rel. Hum. %	87	84	80	79	75	70	74	77	82	86	87	88	81
	EMC %	18	17	17	16	15	14	14	15	17	18	19	19	16.7

TABLE 11:

Location: Plymouth **Annual Average**

Time (GMT)	Temp°C	RH %	EMC %
03	9.0	91	20
09	10.3	91	20
15	12.1	79	16
21	10.1	87	19
Overall Av.	10.4	87	18.7

TABLE 12:
Location: Manchester **Annual Average**

Time (GMT)	Temp°C	RH %	EMC %
03	7.3	87	19
09	8.8	80	17
15	11.4	88	19
21	8.8	81	17
Overall Av.	9.0	84	18

TABLE 13:
Location: Aberdeen **Annual Average**

Time (GMT)	Temp°C	RH %	EMC %
03	5.8	88	18
09	7.7	81	17
15	9.6	74	15
21	6.9	85	18
Overall Av.	7.5	82	17

TABLE 14:
Location: Belfast **Annual Average**

Time (GMT)	Temp°C	RH %	EMC %
03	7.1	89	19
09	8.5	85	19
15	10.9	85	18
21	8.3	86	19
Overall Av.	8.7	86	18.8

TABLE 15:
Location: Dungeness **Annual Average**

Time (GMT)	Temp°C	RH %	EMC %
03	8.7	89	19
09	10.0	85	17
15	11.7	79	16
21	9.7	87	19
Overall Av.	10.0	85	17.8

holding capacity of the air,the higher the temperature the faster the drying.

TABLE 16:

Average air conditions for June, July and August and corresponding equilibrium moisture content for wood.

Locality	Temp°C	RH %	EMC %
Aberdeen	11.8	80	16
Belfast	13.7	81	16
Dungeness	15.4	84	17
London	16.4	73	14
Manchester	14.8	74	15
Plymouth	15.0	84	17

Principles of Air Drying

Except for a few months of the year, the drying elements of sun and wind, together with the influence of rain are unbalanced. Furthermore, it is difficult to control the effect of these elements on wood, but if its final condition and its value are not to be reduced then two fundamental principles must be followed. The first is that air movement throughout a stack must be positive so as to encourage uniform drying, and the second is to ensure the stacks are erected in such a manner as to eliminate costly degrade in the form of distortion, splitting and checking.

It is perhaps elementary to suggest that if two freshly sawn boards are placed one on top of the other they will not dry as well as when sticks are placed between them in order to provide a passage for air circulation. What is not always recognized is, that if two wet boards are allowed to touch and so preclude drying air, they will be in a very suitable condition to encourage the development of unsightly blue sap stain and, in timbers with a low natural resistance to decay, to permit other forms of fungal attack. It does not follow automatically that a seemingly clean, freshly-sawn board is free from fungal spores: species of the fungi *Ganoderma*, which is a wound parasite of trees, including beech, shows only very light coloured

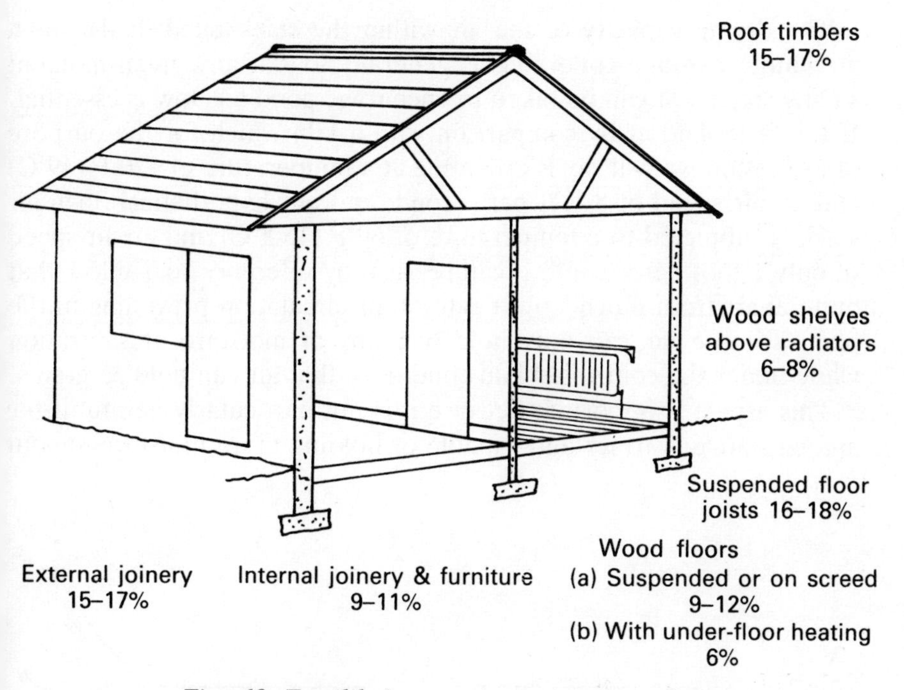

Fig. 69 Equilibrium moisture contents for wood in houses.

mottling of the wood in the early stages of attack and can pass unnoticed when piling, but sluggish air which allows the moisture content of the wood to remain above 20 per cent for a lengthy period will encourage progressive softening of the infected areas and, of course, fungal spores are ever present in the air and ready to settle and develop in suitable wet wood.

If a stack of wet wood is piled with sticks separating the courses or layers of boards so that air can pass through the stack without impediment then the wood will dry. The rate of drying and its uniformity will vary according to a variety of factors including thickness of wood and sticks, the time of year, permeability of the species, siting of the stack, and so on.

Let us take a hypothetical example and assume a man has acquired say, an ash tree which, after felling and conversion has produced 50 cubic feet of boards which he now wants to dry. Green ash weighs about 52 lbs per ft^3 and about 40 lbs air dry, i.e. some 600 lbs of water (or 60 gallons) must be induced to evaporate from the wood, which is a lot of water from a relatively small amount of wood.

The drying capacity of the air within the stack piled in the open air, under average conditions is generally low at any given moment so any steps that can be taken to encourage good air flow is essential. If this is looked at by comparison with a kiln which for the purpose of discussion we will say is operating at a temperature of 120°F (49°C) and an air speed of 8 feet per second, and our hypothetical open air stack is subjected to a temperature of 60°F (15.5°C) and an air speed of only 1 foot per second, it can be seen by reference to Table 8 that quite apart from much slower rate of air circulation prevailing in the open air, the air can only hold 6 grains of moisture at saturation while under the controlled kiln conditions that air can hold 35 grains.

This aspect is important because it is not particularly helpful if the ambient air is warmer and capable of holding more moisture vapour

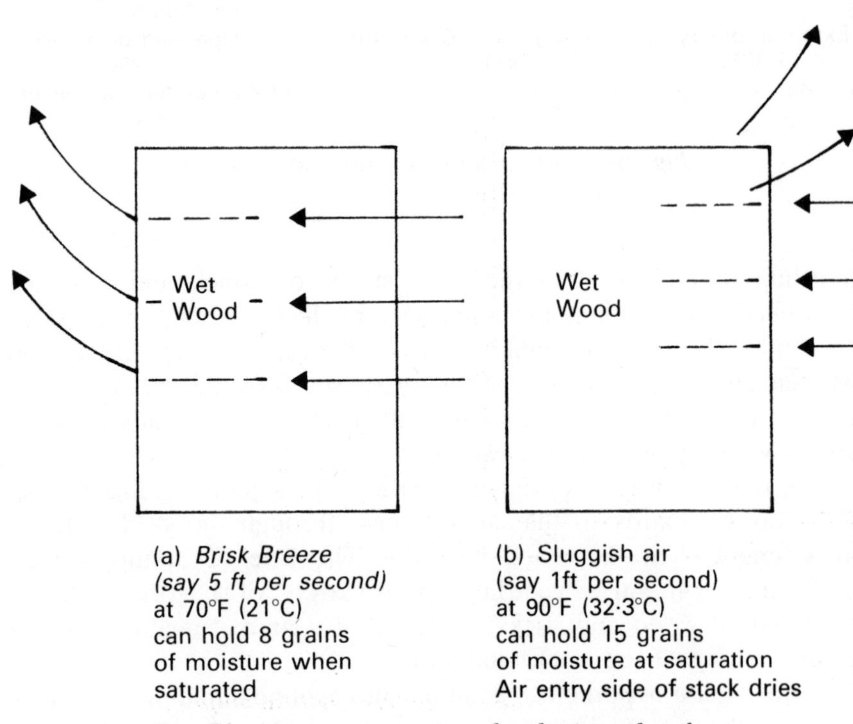

Fig. 70 Air temperature and velocity related to effective air-drying: although air (a) and its temperature cannot pick up as much moisture as warmer air (b), its timber drying capacity is greater due to air velocity

if the circulation is poor. A brisk breeze at 70°F (21°C) is much better in drying terms than sluggish air at 90°F (32.3°C). In a timber merchant's yard, although ground is expensive, as much space as is reasonable will be left between adjacent stacks in order to promote air flow, but the average craftsman may be severely limited in open air drying space. It is essential, therefore, to consider carefully such space as there is before laying down wood for seasoning. Suppose a car port, built on a domestic dwelling appears ideal to stack some wet wood; it has a roof but is otherwise open. This might be perfectly satisfactory for fairly short term drying, but if the wood is to be left there for lengthy periods it is absolutely essential to have regard to

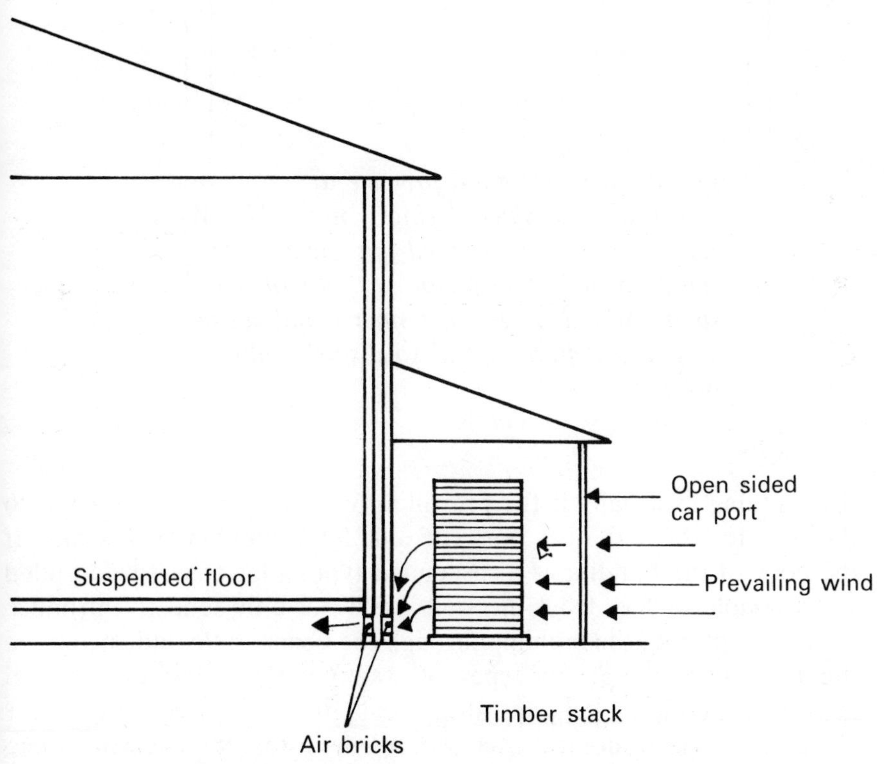

Fig. 71 Care must be taken when attempting to dry wood under a car port. There may be a possibility of moisture vapour being introduced beneath suspended floors via air bricks, or causing mould to develop on rendered walls.

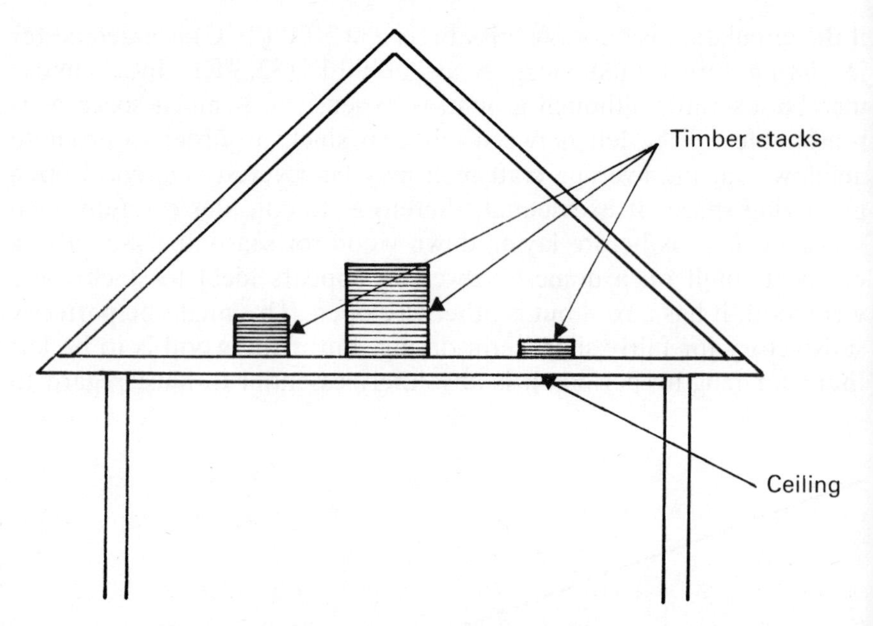

Fig. 72 It is not good practice to store wood in a roof void since drying is negligible. When roof timbers are designed this includes a standard deflection factor of 0.003 of the span. Added weight of timber could cause excessive deflection and so disturb ceilings below.

the building especially if the prevailing wind is generally directed to the wall to which the car port is fixed, for a number of reasons. If the floor of the building is a suspended type, a lot of wet wood piled say a couple of feet from the wall could not only tend to introduce damp air entering the building through the vents, but could encourage the formation of various types of stain on the brickwork, mould spots, for example. It is not always easy to find a suitable spot for air drying under domestic conditions and for that reason some woodworkers may be tempted to put a few pieces of valuable wood in a roof loft where they can be left to dry undisturbed perhaps for a year or more or until required. This is not good practice since drying will be extremely slow, and of course there is a limit to the amount of additional weight a roof truss will carry before deflecting and transmitting the movement to the ceiling below.

Where space is available the following procedures can be followed. The following factors must be known; (1), the average moisture content of the wood; (2), the equilibrium moisture content at the surface of the wood; in this latter respect it means the equilibrium moisture content of the wood corresponding to the air conditions surrounding it.

With kiln drying the average mc is known from the tests and the air conditions are available from the charts at any time, however, it is a simple matter to obtain an average of the moisture content by using a moisture meter, and by reference to Table 7. Assess the potential equilibrium moisture content by assessing the relative humidity of the air. This of course can easily be obtained by using a direct reading RH meter, but in the absence of one it is essential to make an inspired guess. If the wood is drying in the open air, equilibrium moisture contents will vary from winter to summer but logically, the woodworker who wants to know how wet his wood still is in the mid thicknesses in order to gain an idea of how much longer drying it needs, will usually do so in early spring, to assess the drying capabilities of the summer months, or in the autumn. In the UK the driest value for wood stacked in the open air is 15 or 16 per cent moisture content, but during winter this would be a little higher at about 18 per cent mc, so depending on time of year, any of these values could be used as an equilibrium figure.

To find the moisture content in the centre of a plank the following formula may be used,

$$Mc = \frac{3}{2}\left(Ma - E\right) + E$$

Where Mc is the moisture in the centre; Ma is the average moisture content; E the equilibrium moisture content at the surface.

Example: the average moisture content checked by moisture meter is 18 per cent; the time of year is say May so EMC is assumed to be 15 per cent; therefore,

$$\frac{3}{2}\left(18 - 15\right) + 15 = 19.5 \text{ per cent mc at centre thickness}$$

Suppose the same wood was checked later in the year, say in late October and the meter gave an average reading of 20 per cent mc;

by now using an equilibrium value of 18 per cent mc the value at mid-thickness would be 21 per cent mc

A further example of the use of the calculation could be where air-dried wood was removed from the open and was placed in a warm workshop to dry to a more acceptable level. Assume that after a period of further drying the average moisture content had reduced from 18 per cent to 15 per cent according to a meter reading, and the equilibrium mc in the workshop was 10 per cent. By using the calculation the centre thickness would now be 17.5 per cent mc in other words, the wood had dried on average from 18 per cent with 19.5 per cent at the centres in the open air, to 15 per cent average and 17.5 per cent mc in the centres by the warmer atmosphere.

It must be pointed out that such calculations produce only very approximate results, and become even less accurate on thick wood. It may be informative to check the results of a calculation by meter testing freshly cut core surfaces.

If a moisture meter is used intelligently, the results can be as accurate as by the oven dry method but the fact that a moisture gradient exists must be recognised; if the electrodes do not penetrate deeply enough into the wood then the reading may be less than an average. Boards can be cut and the pins inserted in the cut ends, or a special type of hammer electrodes are available for use on dense hardwoods, while longer electrodes are available for thick timber.

A minor irritation with moisture meters is the pin holes left in the wood, and of course, batteries do run down; nevertheless they are invaluable instruments for checking the condition of wood and well worth their moderate cost.

Foundation

Assuming a site has been chosen which will take advantage of prevailing winds it is now essential to ensure the foundation on which the timber will be stacked is firm, level, and well drained. All weeds and debris should be cleared and suitable material placed in position to support the stacked wood. This material could be old railway sleepers or similar dunnage and should be high enough to promote a brisk air movement at the base of the stack and there should be a slight fall from the front bearer to the back of the stack to allow rain water to drain off the stacked wood.

Since the spacing of the base bearers will have an effect on the straightness or otherwise of the stacked planks, they should be set at distances determined mainly by the distortion tendencies of the wood species to be piled and their thicknesses. The sticks used for separating the courses of timber should of course be placed in alignment directly over a base bearer. Table 17 gives a guide to bearer and stick spacings.

Once the bases are down and the pile is started it is bad practice to attempt to put additional sticks in each course because the base bearers have been set too far apart. If this is done then as the wood dries its weight will cause the boards below the extra, unsupported sticks, to bend. The only time extra sticks are needed is when the odd board on occasions will not meet the nearest stick at one end. A short stick under this end will obviously support it but a few additional short pieces ought to be placed immediately under this stick to level out the weight imposed as the wood dries.

It should be noted that Table 17 refers only to stick spacings, not the size of the sticks; in other words, while it would be satisfactory for thick oak, for example, to have sticks set 4 ft (1.2m) apart, it would

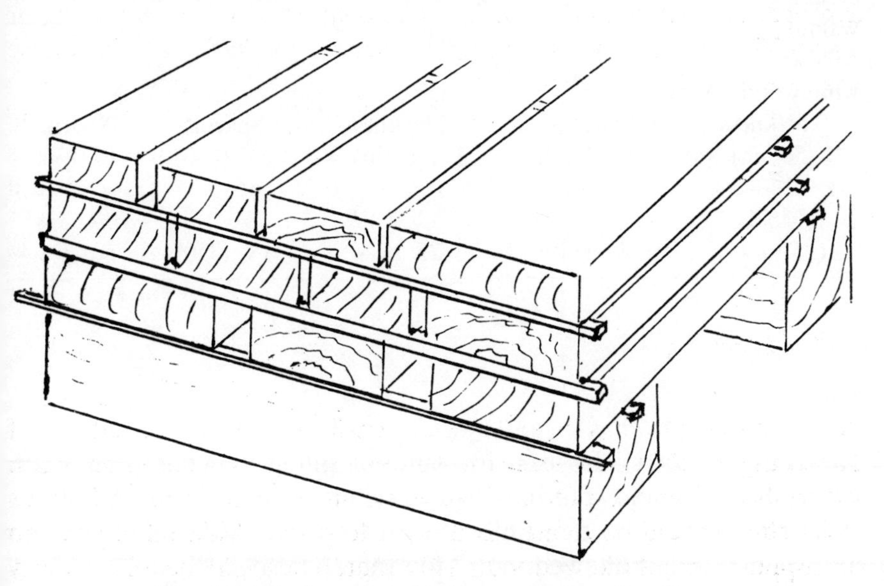

Fig. 73 Timber should be stacked well off the ground on suitable bearers; sticks should be kept flush at front and be laid in alignment over bearers

be entirely wrong to use thick sticks. On the one hand we are thinking of encouraging air flow by reducing as much impediment between the courses, compatible with the wood's ability to resist distortion, while on the other hand we want to control as far as possible the rate of drying compatible with the wood's ability to resist checking.

TABLE 17: SPACING OF PILE BEARERS AND STICKS USED IN AIR DRYING RELATED TO DISTORTION CHARACTERSTICS

Timbers with low distortion tendencies		**Timbers with appreciable distortion tendencies**		**Thin timber prone to distortion**
alder		ash		elm
basswood		beech		beech
birch, paper		birch, Eur.		gum
birch, yellow		cherry		pear etc.
lime		chestnut, horse		
magnolia		chestnut, sweet		
oak		dogwood		
sycamore		elm, Eur.		
walnut		gum, red		
willow		pear		
whitewood, Amer.		poplar		
Thickness	**Spacing**	**Thickness**	**Spacing**	**Spacing**
2″ (50mm) +	48″ (1220)	2″ (50mm) +	24″ (600)	12″ (300)
1″ to $1\frac{3}{4}$″ (25 to 44)	24″ (600)	1″ to $1\frac{3}{4}$″ (25 to 44)	18″ (450)	(*mm in brackets*)

Sticks

The thickness of sticks used for outdoor piling depends very much on the checking and splitting characteristics of the wood in question. Oak and beech, and other timbers with large rays are liable to severe surface checking if allowed to dry too rapidly, especially during warm weather if this coincides with the first few months of exposure. For valuable oak, thin sticks of $\frac{1}{2}$ inch (13mm) or $\frac{5}{8}$ inch (15mm) are best, while hardwoods with low checking tendencies, (Table 6) can be piled with 1 inch (25mm) sticks. Green softwoods on the whole can be piled with 1 inch (25mm) sticks except green Douglas fir and western larch, both of which tend to check; thinner sticks, say $\frac{3}{4}$ inch (19mm) are best for these timbers.

It is important that the thickness of all sticks is uniform, dressed to size off the saw or surfacer and, as far as possible, the sticks should be of square section. There are two reasons for this: if the width of the stick varies slightly from the thickness there is the probability that some sticks will be turned over when placed in position and thereby encourage distortion as the boards dry. If, on the other hand they are somewhat wider than they are thick, then there is more chance of stick marks developing.

The question of width in sticks depends a lot on circumstances; a craftsman may want to use offcuts from a variety of species. Under these circumstances it should be borne in mind that if such sticks contain a lot of sapwood this can contribute to sap staining of the wood and may attract beetle pests. Strips of chipboard that might be

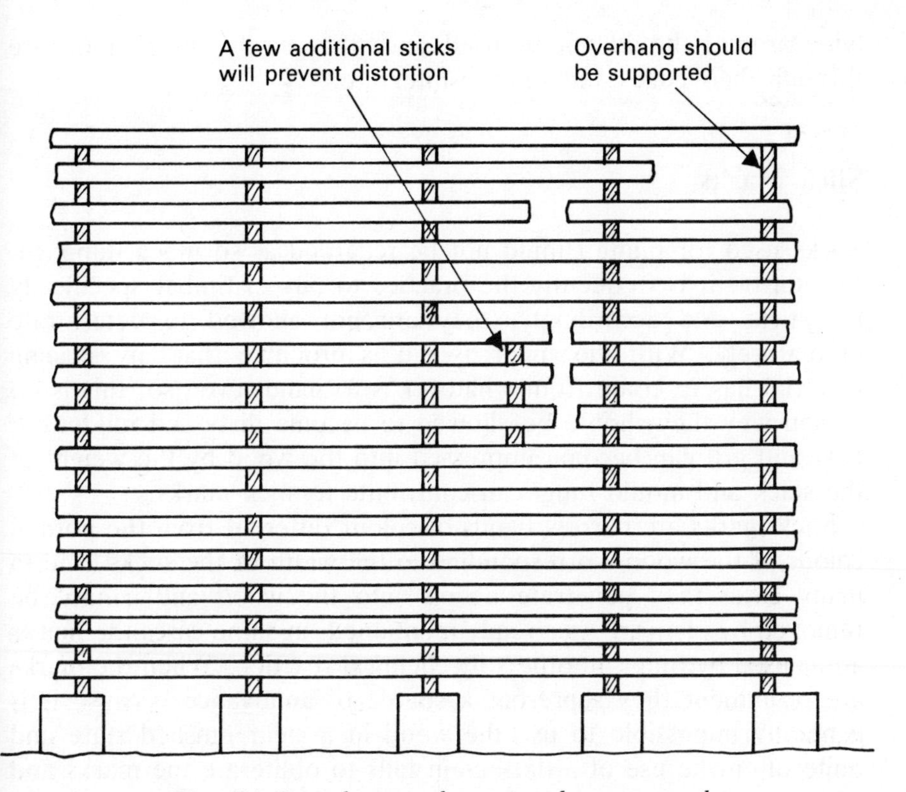

Fig. 74 Boards must be properly supported

Fig. 75 Poor sticking encourages warp

lying around should not be used since they easily absorb moisture through their edges and soon disintegrate.

Stick Marks

Sticks used for piling should not be regarded as so much dunnage. In industry it is frequently the practice to buy in timber specifically for sticks and accordingly, they are not allowed to deteriorate prematurely. With the small user it is probable that any spacing material has to come from whatever is to hand; even so, this is no reason why they should be allowed to become dirty and mildewed. Dirt and grit can become impressed into the wood by the weight of the stack and mould fungi can contribute to stick marks.

Stick marks are narrow bands of colour different from the normal colour of the wood, corresponding to the width of the sticks, and in many cases they penetrate deeply into the wood and cannot be removed by planing. As already mentioned, in some cases the marks are caused by fungi, in others by chemical reaction. When the marks are prominent they represent a source of annoyance because it is generally impossible to use the wood in a clear finished state and quite often the use of a dark stain fails to obliterate the marks and in fact tends to enhance them.

When wood dries, its original colour changes somewhat; light-coloured woods tend to darken and this is brought about by colouring pigments in the wood migrating to the surface as moisture is extracted. In the normal course of events, as the wood dries, the effect of air at ambient temperature combined with humidity brings about a chemical change in the wood surface layers which tend to darken. Under the sticks however, the reaction is different: the drying rate is slower as is the migration of pigments and various salts, and the result is a difference of colour, often to the full thickness of the wood.

With some woods, enzymic reaction takes place; in others, and under different exposure conditions, oxidation or hydrolysis is thought to be responsible for stick marks, but whatever the reason, certain woods containing a high degree of tannin, like oak, sweet chestnut, and western red cedar will react with other woods containing infiltrates, iroko is a notable example. In some cases, leaching out of certain

Fig. 76 Air drying Maritime pine in Portugal. It will be noted that end racking (beyond stacked timber) precedes stacking proper. Photograph, courtesy Forest Services Sawmill, Marinha Grande, Portugal

pigments can cause stick marks; idigbo from Africa is a case in point; under moist conditions it leaches a yellowish colour and the most moist areas, of boards stacked in the open to dry, is under the sticks.

Softwood sticks, say of spruce or pine for preference, are best, but if, for one reason or another, they are unavailable and stick marks must be avoided at all costs, the potential formation of these can be reduced and often eliminated by end racking the boards for a week or two before placing them in stick. The end racking should allow the surfaces of the boards to dry out and by doing this, any migration of chemicals is relatively minimal.

It is preferable that sticks should be dry at the time of piling to avoid high moisture levels and accompanying marks and staining at the stick/board interface.

Another way is to use specially prepared sticks whose contact surfaces have been reduced (Fig 77). Well prepared sticks of any type are expensive and ought to be regarded and treated as one would regard a section of moulding, and although it would add to the cost, it would also add to the value if sticks were moulded to shape as in the sketch. Such sticks could be kept in a safe place and used when expensive hardwoods were being air dried.

Methods of Piling

Much timber for air drying today consists of unit packages piled with the aid of a fork lift truck, but the craftsmen who dries his own stock usually does this by hand. There are a number of ways in which to pile and a number of variations to each method. Whatever the method chosen, it should be designed with the following factors in mind. Air flow should be as unrestricted as possible; protection from excessive sun and rain must be provided; and warping tendencies must be restrained.

Assuming the base foundation has been prepared properly, the wood can be laid down with each course separated by appropriate sticks. There are two ways in which to regard the front of the stack: in one, as each course is laid down, its boards are brought forward very slightly instead of being kept flush; this results eventually in an overhanging protection of the front ends of the boards from too rapid drying, especially from the sun. In the alternative method the front is kept flush, but with some hardwoods, notably oak, the front sticks

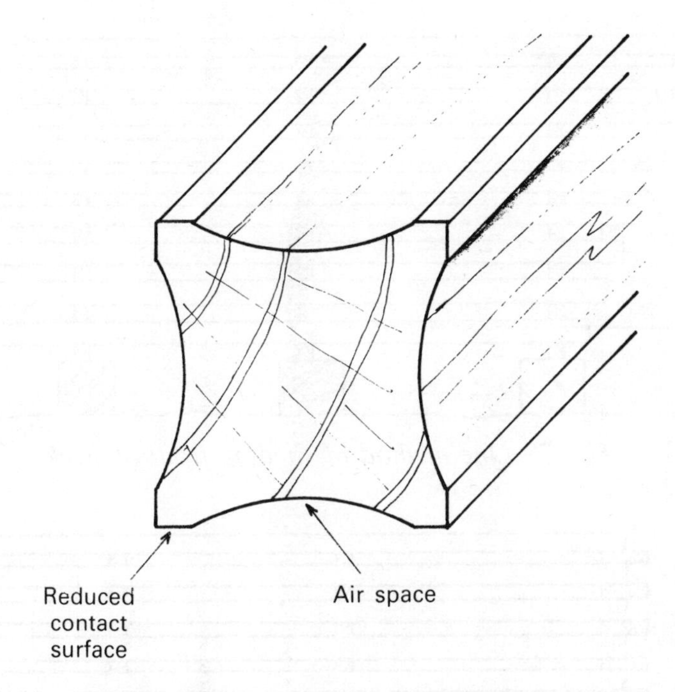

Fig. 77 Sticks moulded to profile to give reduced contact between boards and help reduce stick marks.

used are wider than the others and half the width is allowed to project and so offer a degree of shading to the ends of the boards.

If this is the general method of piling adopted and only the front of the stack is made flush, the back of the stack will consist of gaps and board ends needing support. With random length, random width boards, it is better to keep both ends of the stack flush, selecting out the lengths to suit, and making sure, as far as possible, that the longest lengths are used on the outsides of each course, filling in the middles with shorter stock. It is often possible to make the length of the stack correspond entirely to the length of the longest boards.

When square edged boards are being piled, and these are at a high level of moisture content, it often helps to build a flue or chimney in the centre of the stack, in other words a gap of several inches is left in the middle of the first course laid down and continued in each succeeding course, but progressively being lessened in width so as to produce a gradually narrowing space from bottom to top of the pile.

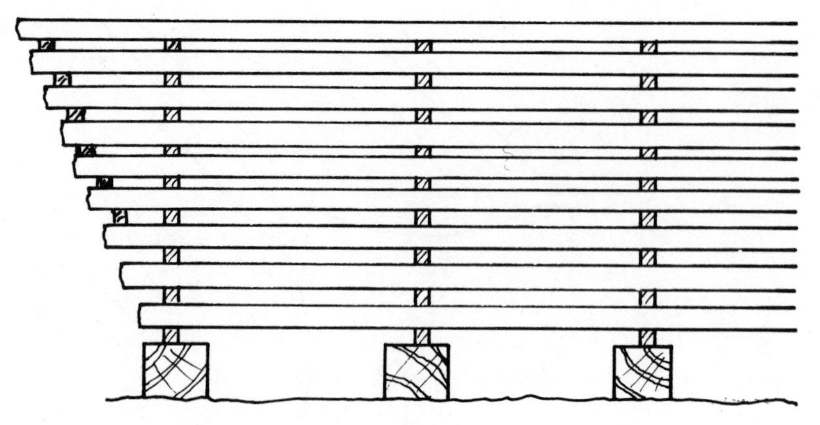

Fig. 78 One method of shading front of stack.

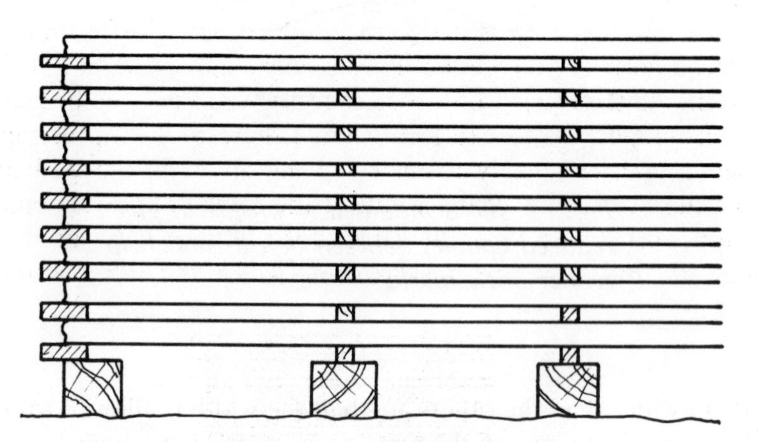

Fig. 79 Wider sticks at front of stack allowed to project will offer shade from sun.

In air drying wet wood, there is a fairly well established pattern of moisture removal from the wood especially in the first few months of exposure and in a given stack this results in six different areas of rate of evaporation. Firstly, the top one-third dries better than the middle-third, with the lower-third drying slower. There is a varation, too, between the outsides and centre of each third, the outsides of the stack generally drying in advance of the centres. While there is a narrowing of the gap between top and bottom of the pile long term, there is still a difference in moisture content uniformity. Obviously, no stack can be expected to dry uniformly throughout, but if air circulation can be evened out in the stack then the drying time at least will be shortened.

The graph Fig 83 indicates the typical variation in moisture reduction in timber piled in the open. It was extracted from an American publication *Air Drying of Lumber, Agriculture Handbook No 402.* (US Dept. of Agriculture; Forest Service; 1971) and although American air conditions vary somewhat from our own, the drying pattern is similar. The white pine used in the field test is a relatively easy drying wood but if the graph is considered it can be seen that the drying rate of the centre boards in the middle and lower portions of the pile lagged well behind the entire top third drying rate and although after four and a half months exposure when the gap began to narrow, there was still a difference of some 10 per cent of moisture content between the driest and wettest portions of the pile, and the difference could have been greater had the timber been a refractory hardwood.

When square edged boards are set out in each course, they should be placed fairly close together leaving only a gap of about 1" (25mm). In air drying they ought not to be touching edge to edge because this will encourage sap stain to form should there be a strip of sapwood present. By leaving only a small gap the air is encouraged to flow across the boards and if the width of the stack is kept to something like 6ft (2m) this is generally satisfactory. However, in poor drying conditions, or where the stack is made wider, the introduction of a

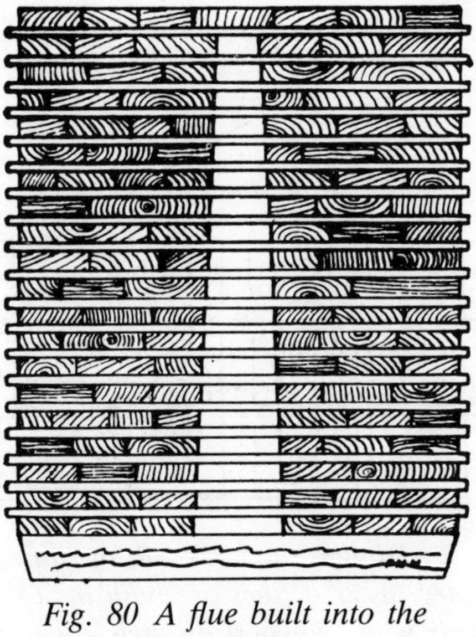

Fig. 80 A flue built into the stack will promote air circulation

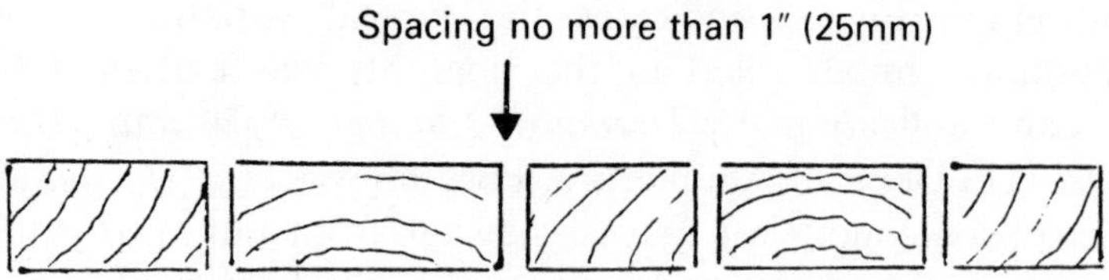

Spacing no more than 1" (25mm)

Fig. 81 Random width stock should be piled close together to encourage air flow over each course

flue will greatly help to promote more uniform drying. In the normal course of events, warm air will tend to be drawn up a flue space and under some circumstances in a yard this will happen in a timber pile. More often than not, however, the air within a stack may get very close to dew point at times, and cold, moisture laden air tends now to fall and be drawn through the outsides of the lower portion of the stack. A flue built in the stack now acts in reverse and assists in expelling some of the moist air at the lower levels. In some instances, say with softwoods (except Douglas fir and larch), and with easy drying hardwoods like alder, lime, poplar and willow etc, if these are sawn all to one width, i.e. dimension stock, their drying can be hastened by leaving an appreciable gap between each board; in other words, by making a flue between each pair of boards in each course.

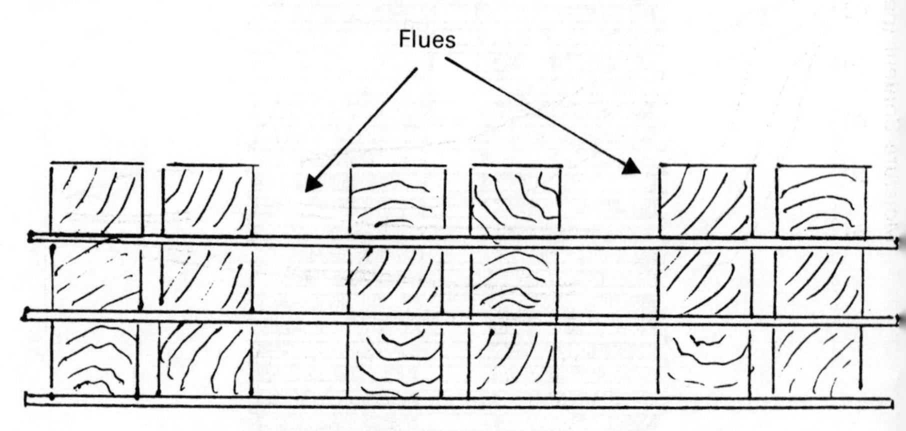

Fig. 82 Dimension stock, all of same width, will dry faster if flues are built in. (Note: Boards with sapwood on their edges should not touch in order to reduce blue sap stain)

Box Piling

There is a variation of stacking known as box piling which generally should be avoided. It is sometimes used for air drying fairly short lengths and narrow width stock and is applied by placing a row of boards on a base, with a fair-sized gap between each board, and then placing each succeeding row at right angles to the row beneath. If the timber is all of the same length the pile is kept square, if not, then the longer ends are allowed to project unsupported.

Despite the wide gaps in the pile the wood does not dry uniformly and because of the number of contact areas each board is liable to develop bad staining. The only merit would appear to be that speed of piling is achieved and no sticks are used.

Box piling is sometimes used for square section timber intended for furniture legs and similar, but the method requires qualification.

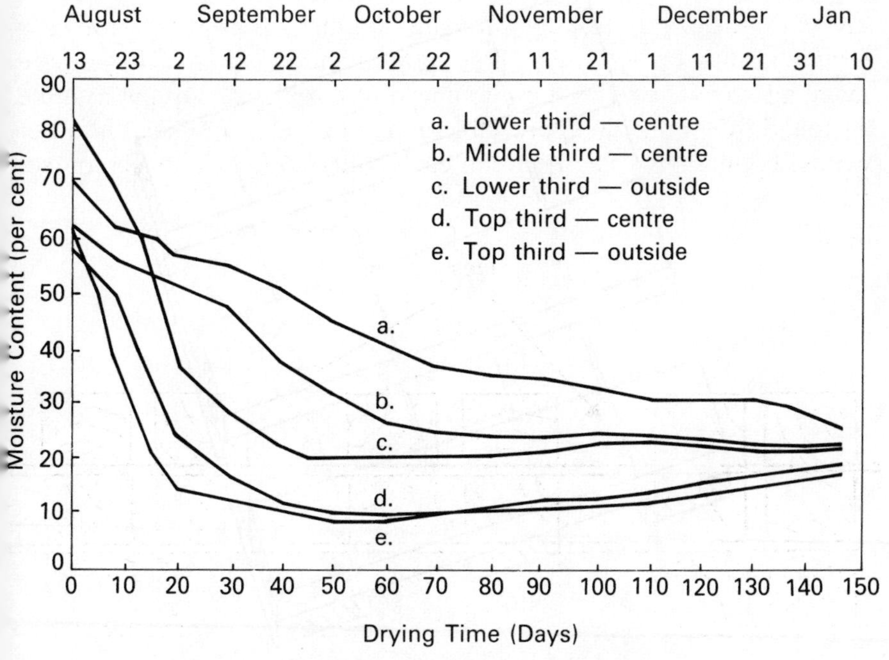

*Fig. 83 Moisture content time graph for top, middle, and lower portions of a hand-built pile of 1 inch (25mm) western white pine (*Pinus Monticola*)*

If squares are sawn from fairly dry and small section timber then box piling is not necessarily bad practice because largely, the degree of earlier surface drying has reduced the liability for discolouration to occur under contact of one piece with another. It is usually safer to place square sections in stick if they require further drying.

Log Sawn

Where a log has been sawn through and through, or with one square edge, it is desirable for the wood to be placed in stick in reconstituted log form, i.e. as it fell from the saw. By this means, colour and grain matching is easier when the wood is eventually used and the pile, being narrow dries more readily. With small diameter logs such as holly, boxwood, and dogwood, all of which tend to split badly when

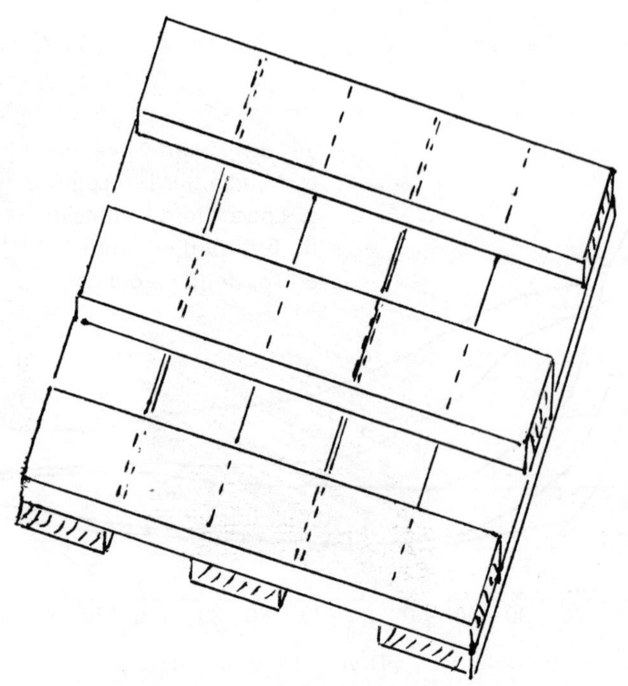

Fig. 84 Box piling. Although this method allows good air circulation through the stack, drying is non-uniform due to the many contact surfaces which also encourage staining

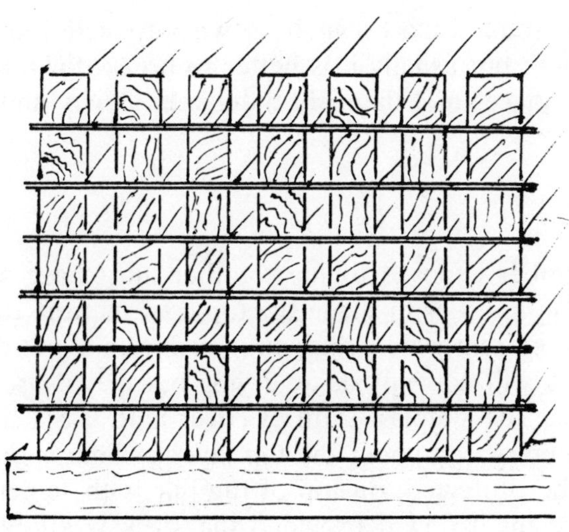

Fig. 85 Squares should be placed in stick to season when freshly sawn

Fig. 86 Sawn logs should be piled as falling from the saw. The wood must be kept clear of ground and wedged if necessary.

dried in the round, these can be sawn into half rounds and then placed in stick, but usually it is better to break down the wood into suitable size pieces and then dry them very slowly under cover.

Pile Cover

A pile of timber exposed to the weather must have some form of top protection in order to shade the top courses from the heat of the sun and to shed rain water, since too much rain allowed to percolate into the stack will not only delay drying but may cause water marks on the wood. The simplest form of cover is to use a sheet of heavy duty polythene placed under the top layer of boards or, if the latter are also to be protected, on top of the pile with something like old timber to weight it down. Provided the stack was built with a fall, rain water will be shed quite satisfactorily although the polythene will have only limited life. If this form of cover is used, the sheeting should not be draped over the sides of the stack since not only would this interfere with air circulation, but during warm weather could encourage condensation under the sheeting to the extent that black spots due to mould fungi could form on the wood.

A better roof economically is to use corrugated plastic sheets weighted down with dunnage. The sheets should be allowed to project slightly beyond the front and back of the pile so as to shed water away from the wood. Old galvanised iron sheeting should be avoided for roofing since rain water gaining access to the wood could be contaminated with iron salts. Timbers with a high tannin content like oak, chestnut, and western red cedar in particular, could be heavily stained by contact with water dripping into the pile. Severe wetting of some woods from rain water can likewise cause chemical stains to develop due to a tendency for soluble pigments in the wood to be released; idigbo and merbau are examples, and it is of common occurrence with sweet chestnut.

End Racking

End racking, or end piling is generally recommended as a means of retaining the natural light colour of sycamore. The method ensures the surfaces of the boards are dried out rapidly thus reducing the migration of pigments to the surface. End racking in its fullest sense implies that a suitable wooden frame or rack, substantially built, is available, against which, on opposing sides, the boards are placed, usually on edge. Alternatively, the boards are sometimes placed flat-on, each side of the frame, with successive layers, separated with sticks being built on either side, which in effect means literally, two stacks tilted obliquely against a frame. The craftsman who wishes to practice end racking with only a moderate amount of boards can achieve the same result by placing the first board at an angle against a suitable support and following up by placing successive boards so that only the top end of each touches the support, the remainder of the length being angled away so as to leave a space from the board beneath.

End racking of sycamore and maple can help considerably in holding the nice white colour of wood as it falls from the saw, but it must be remembered that the natural colour of all wood is transient and eventually, after exposure to light, colour changes will take place although it may take a long time to do so.

Air Drying Costs

Costs incurred in holding timber during air drying will vary considerably from one situation to another. The craftsman storing small quantities of wood will probably be concerned with drying degrade, that is to say with the actual amount of usable dried wood recovered, and with shrinkage loss, measured against the purchase price. On the other hand, the small manufacturer operating on borrowed money, will be concerned with drying time, since the longer his stock is seasoning and unproductive the longer it will take to discharge his monetary liablilities.

There are many cost factors involved, many of them uncontrollable. Those that can be controlled, such as drying time and drying degrade, are nevertheless variable factors since the man drying green softwoods will have shorter drying times and lower potential degrade than the man drying green hardwoods.

The use of stick thickness, satisfactory to the best drying rate of the timber and the restriction of stack width, will contribute to a shorter drying time; and in some cases consideration can be given to trimming the edges of through and through stock at the time of conversion since there is no point in seasoning what will ultimately be discarded as waste.

The full costs of air drying will vary from site to site but a good guide to basic capital costs is that devised by D.D. Johnston of the Princes Risborough Laboratory who evolved the following equation.

$$\frac{t}{k}\left[(C + L + T)\,r + C(x + y) + T(z)\right]$$

Where t = mean drying time in years
k = yard capacity in cubic meters
C = capital cost of establishing a drying yard (including roadways, pile covers, foundations, sticks etc.)
L = land value
T = value of timber held for drying
r = current interest rate (value/100)
x = depreciation (value/100)
y = maintenance (value/100)
z = insurance and office expense (value/100)

To give an example, assume hypothetical situation where,
t = 9 months (0.75 year)
k = $500m^3$
C = £7000
L = £30,000
T = £50,000
r = 0.09
x = 0.20
y = 0.14
z = 0.02

The equation is now,

$$\frac{0.75}{500}\left[(7000 + 30{,}000 + 50{,}000)\,0.09 + 7000(0.20 + 0.14) + 50{,}000\,(0.02)\right]$$

$= 0.0015\ (7830 + 2380 + 1000)$
$= 0.0015 \times 11210 = £16.80 \text{ per } m^3$

The resultant cost does not include the cost of labour, nor does it take into account any loss arising from seasoning degrade. The example given is purely hypothetical and it will be noted that the value of the timber held for seasoning would, in practice, represent perhaps not more than half of what the value would be were the yard full; it is obvious therefore that seasoning costs can vary considerably according to all conditions of the site.

Labour and maintenance costs are variable quantities depending upon whether the yard is mechanised, and to what extent, if at all. A yard with little mechanical aid will generally be more labour intensive than a similar yard partially mechanised, but on the other hand maintenance costs will be higher as a rule, in the latter yard. Stick costs and level of replacement will vary depending on the types of timber held and the general thickness and grade of the sticks used.

7
Drying Defects

A defect in wood is any irregularity or imperfection that reduces its volume or quality and in this sense abnormal growth, knots, and insect or fungal attack are all defects. However, defects caused through poor seasoning practice may more properly be called degrade since the effect of this is to reduce the initial quality of the green timber by at least a grade. Prime or First Quality commercially graded timber is relatively free from visual defect, while Clears is ostensibly completely free. At the lower end of the grading system, Merchantable means what it says, i.e. saleable or in demand for a particular purpose.

In terms of drying degrade, if a top grade of wood is spoiled by poor drying it drops in quality and value by one or more grades; but if a low grade parcel suffers through drying it becomes useless. The term low, in this context refers to the fact that in commerical grading of wood, as the grades fall from the top, more natural defects are permitted by the grading rules. Many craftsmen prefer these falling grades because of the aesthetic appeal of sound knots and abnormal growth patterns. A packing case maker will want relatively low grade wood for economic reasons; a craftsman sees the wood in a different light and for many of his purposes, the lower grade is more valuable than the higher one, but only if it is dried carefully.

Distortion

Wood may warp, bow or twist during drying although this type of defect is not necessarily caused by poor drying techniques. In kiln drying for example, the schedule used may be suitable for the species in question, but the parcel may contain a proportion of abnormal wood, say with spiral grain, or the wood may have been resawn from case hardened stock. In good kiln drying techniques, however, the kiln operator making his periodic visual inspections can often apply

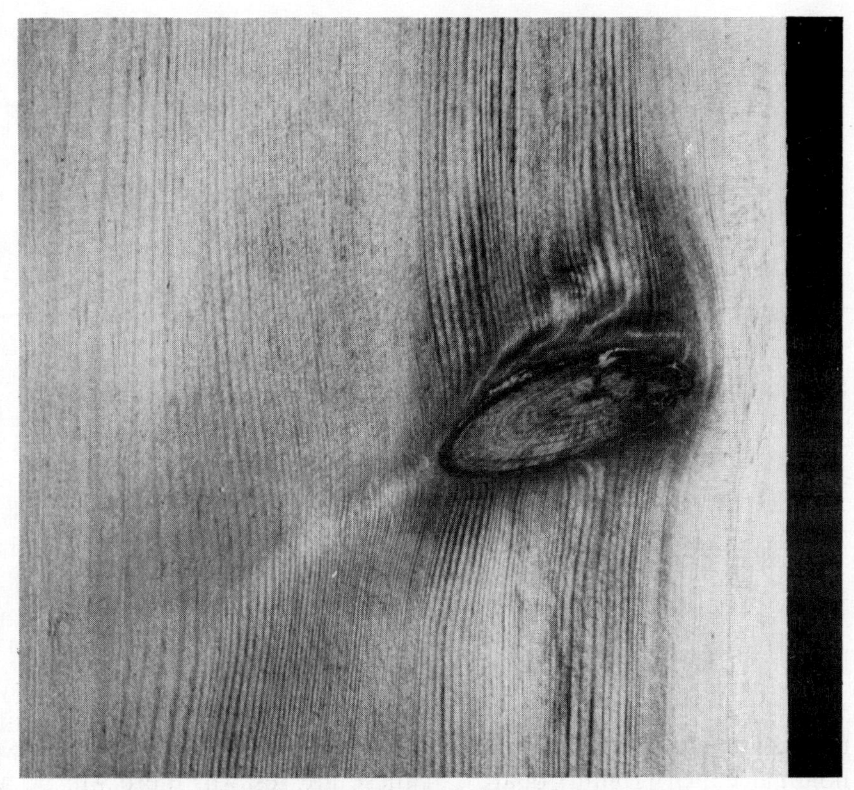

Fig. 87 Knots are strength limiting defects, but such wood frequently has aesthetic appeal. Care is needed in seasoning to prevent knots loosening

a reconditioning schedule to relieve stressed wood but it is difficult to straighten out crooked and bowed boards.

The woodworker setting out his own timber to air dry has limited control over distortion tendencies although in certain cases there are steps he can take to reduce them. Where small quantities are concerned, a small wooden frame with wedges, or a cramping device can be used to tighten up the drying wood both vertically and horizontally.

A question which sometimes arises is when timber should be resawn. Suppose thin boards are needed for drawer sides and backs; if this is done initially from the green state, then drying in the open air or kiln chamber is straight forward since the moisture content is well distributed through each piece; but suppose some $1\frac{1}{4}$ inch (32mm)

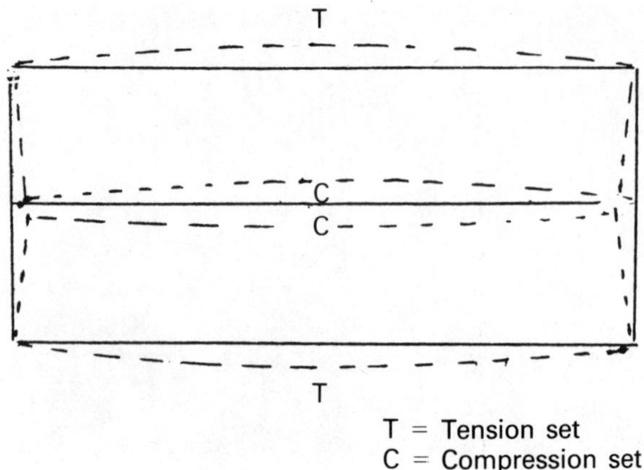

Fig. 88 If kiln dried wood is deep sawn it can be expected to react by cupping as per the dotted lines.

flat sawn and dried material is to be centre deeped to produce nominal $\frac{5}{8}$ inch (16mm) boards. If these are resawn, there will be a tendency for most pieces to cup towards the saw. Stresses set up during drying permanently strain the surfaces of the wood. Generally, these strains are balanced and the wood remains flat. Deep cutting creates an imbalance of surface strains resulting in the cupping. From this it will be seen to be inadvisable to deep cut dried wood. As a guide, if wood has to be deep cut, then it should be cut before much drying has occurred.

When wood is stressed it becomes distorted or strained. Strain produced by short-term stress below the proportional limit substantially disappears when the load is released. This strain is called the elastic strain. Stress beyond the proportional limit, or stress below the proportional limit applied long term, produces some strain that does not disappear when the load is released. This permanent strain is called set.

Thus, when drying wood we have to consider stress, strain, and set. Basically, stress is a mutual force per unit area between adjacent surfaces; strain is a change of dimension, and set is a permanent strain.

In studies designed to measure drying stresses, green boards are marked off at points where test sections will ultimately be taken out. The wood is placed in a kiln and periodically during the drying test sections are removed. Each test piece is measured lengthwise by means of a micrometer and the section is then cut into ten slices. Each slice is again measured and the difference in the two measurements gives the elastic strain. From these strains, actual stresses can be calculated.

Drying stresses that have not been kept under control contribute to, or are the cause of, surface and end checking, splitting, collapse, internal honeycomb checking, and severe case hardening. In kiln drying, adequate control measures are present, but in air drying control is only possible by common sense applied to the method of piling, the protection of the pile from excessive heat, and a recognition of those timbers and thicknesses which need special care.

TABLE 18: WARPING TENDENCIES

	High	**Moderate**
(a)	*Hardwoods*	
	Ash,	Lime
	Basswood	Birch, yellow
	Beech	Elm, rock, Amer.
	Birch	Hackberry
	Cherry, Eur.	Hickory
	Cottonwood	Locust
	Elm, Eur. & Amer.	Magnolia, southern
	Holly	Maple
	Pear	Pecan
	Plane	
	Robinia	
	Sycamore, Amer.	
	Tupelo	
(b)	*Softwoods*	
		Cedar, Atlantic, Lebanon, deodar
		Larch
		Sitka spruce, home grown

Fig. 89 Ayan machined for a lorry floor. This wood has a tendency to warp in seasoning but the outer board, at right, has developed an excessive tension stress due to abnormal growth

Shrinkage and Stress

When green timber commences to dry the outer surfaces of the boards will attempt to shrink but will be restrained by the wetter cores. This results in the surface zones becoming stressed in tension and since a tension stress is balanced by a compression stress, the inner zones are held in compression. If drying proceeds in a uniform manner and the inner zones shrink in unison with the outer zones, then the stresses either reverse themselves or the wood is free from stress once it is dry. This is the ideal; unfortunately wood cannot always be persuaded to behave this way.

In the contolled air conditions prevailing in kilns and drying chambers it is common practice to work to 'safe' drying schedules in order to avoid excessive stress forming in the wood or to take appropriate steps to relieve bad stresses once they have formed. Without close control of the drying elements, as is the case with air drying, it becomes virtually impossible to do other than hold the stresses as they occur, which means that the finally dried wood will have suffered very little degrade, but it will have some residual stress and, to a minor degree will be case hardened.

Wood, if held in restraint, when changing dimension will undergo the phenomenon known as "set". For example, during drying the outer zones of the wood are attempting to shrink, but are restrained by the still wet non-shrinking cores. If drying proceeds too rapidly, the outer zones are larger in dimension than they should be at that particular moisture content and they become permanently stretched. This stretching or "tension set" is commonly known as "Case Hardening". It must be emphasised that the wood is not case hardened

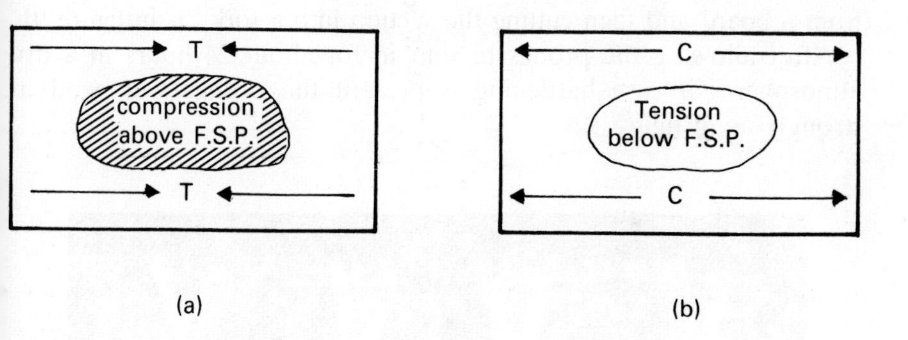

Fig. 90 Normal stress pattern (a) When green wood commences to dry, its outer zone, below fibre saturation point (FSP) attempts to shrink but is held by the lower zones. The surfaces become stressed in tension (T), with the inner zones stressed in compression (c). (b) As drying proceeds deeper in the wood, the stresses tend to reverse. If drying is too severe, the surfaces are now liable to become set in tension and this can lead to case hardening and/or surface checking. Continued harsh drying may produce internal checking.

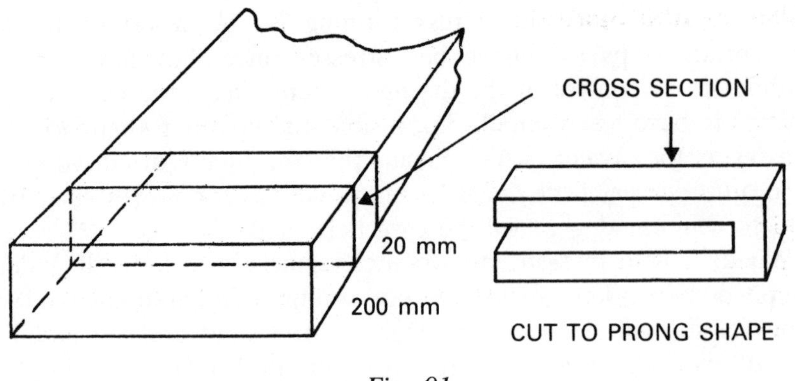

Fig. 91.

in the engineering sense. Generally, the term is widely used in relation to wood but is misleading and to be deprecated.

Timber may be tested for case hardening by cutting a cross section from a board and then cutting the section into a fork, as in figure 91.

After allowing the prongs to remain for about 24 hours in a dry atmosphere, if case hardening is present the prongs will bend in strongly as in figure 92.

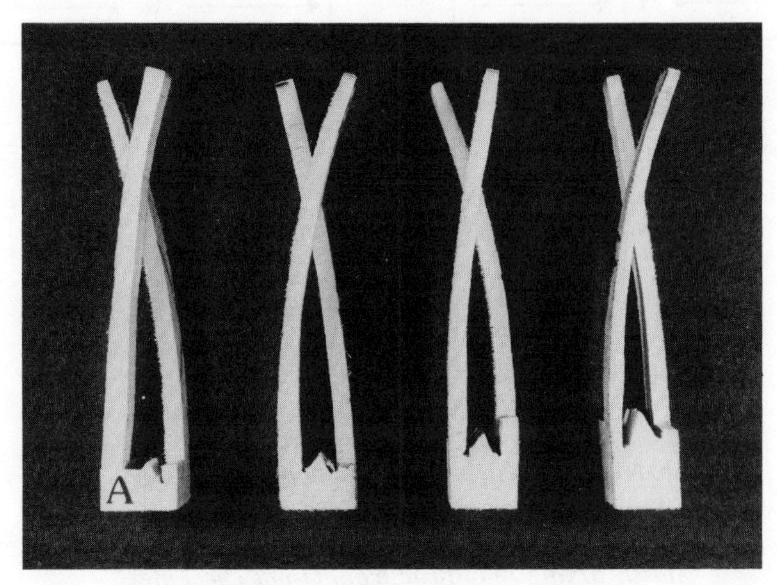

Fig. 92 Drying stresses (case hardening)
Photograph, courtesy of Peter Y.S. Chen.

Checking and Splitting

The problem when air drying green wood is how to prevent splits and checks forming.

Some wood species tend to split and check more than others but woods with large rays and fairly straight grain are the worst in this respect, the rays tending to cleave the wood as it shrinks around them. Table 6 gives a guide to the surface checking proclivities of some hardwoods.

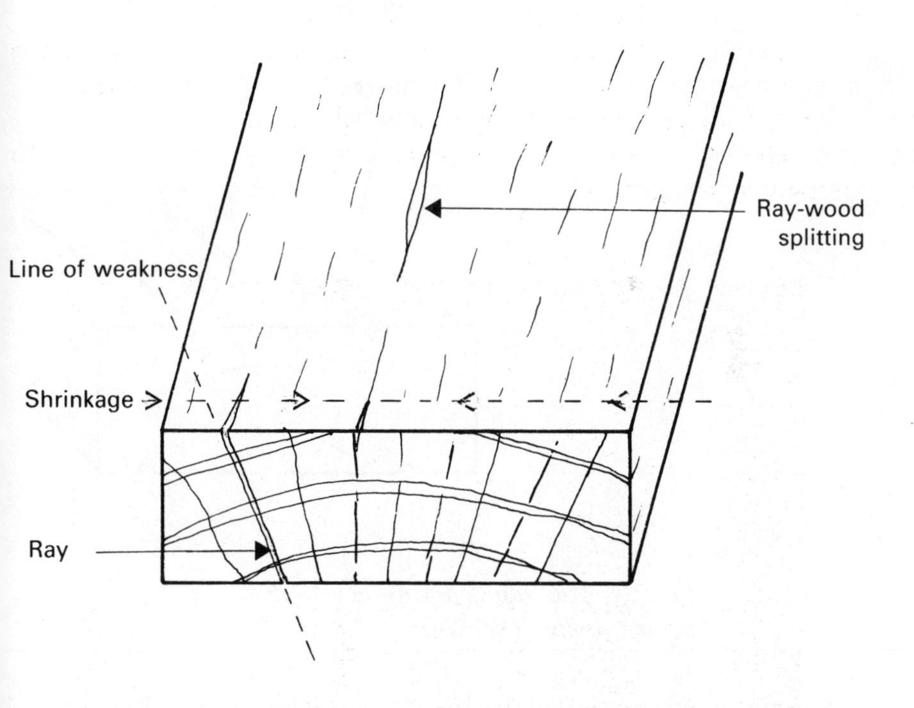

Fig. 93 Cleaving action of wood in area of large rays during seasoning. The ability of wood to split readily along the rays is used to advantage when producing cleft oak and chestnut fencing material and roofing shingles.

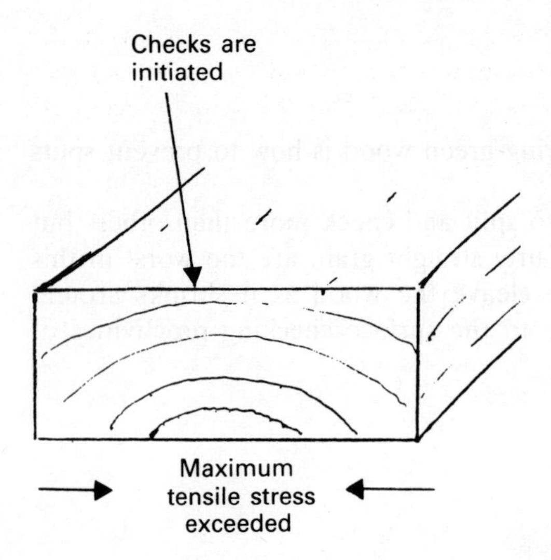

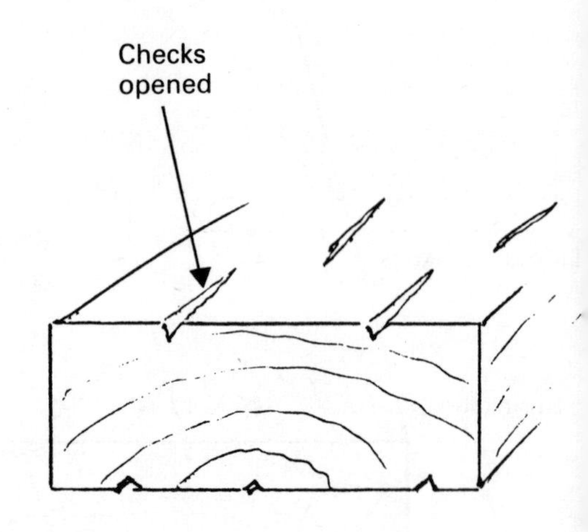

Fig. 94 Too rapid drying at wood surface invites checking

When the moisture content in the surface zones of wood is reduced to a value so low that stresses exceed the maximum tensile strength of the wood perpendicular to the grain, checks and splits occur. Tensile stresses are very great, tending to pull the wood apart, the danger level being reached very early in the drying process. In kiln drying, provided the conditions in the kiln are compatible with the recommended schedule, the stresses are not allowed to become excessive, but in air drying green wood it is often difficult to recognise the fact that if the drying conditions in the early stages are incompatible with the wood's ability to accept, then checks and splits are initiated but are not visible to the naked eye. The fact that the maximum tensile stress has been exceeded will not become evident until later on; in fact surface checks may well remain hidden until after the wood has been put to use, when atmospheric humidity changes cause them to open.

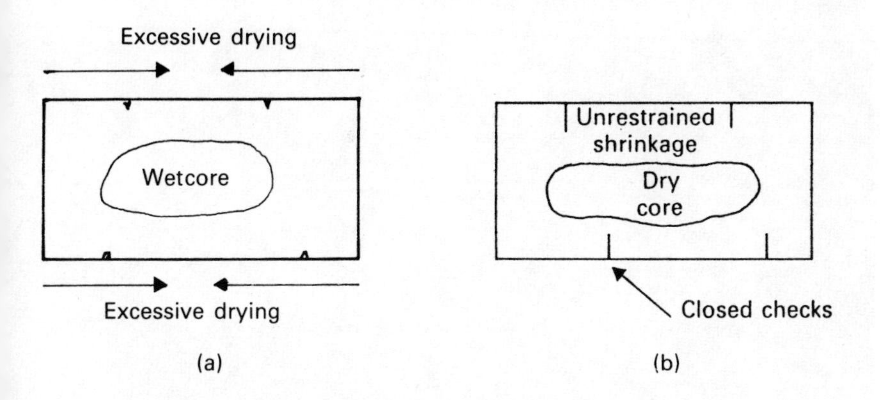

Fig. 95 (a) Splitting and checking of wet wood is the result of severe drying conditions. Shrinkage (S) at the surfaces will exceed the wood's capacity to resist and areas of weakness will occur resulting in checks.
(b) Later, when the lower zones are able to shrink without restraint, the weakened areas will close up, producing closed checks.

Internal Checking

Internal checking or honeycombing in its most common form occurs along the wood rays in tangentially sawn hardwoods. It is sometimes found in radially sawn Douglas fir and a few other woods in the form

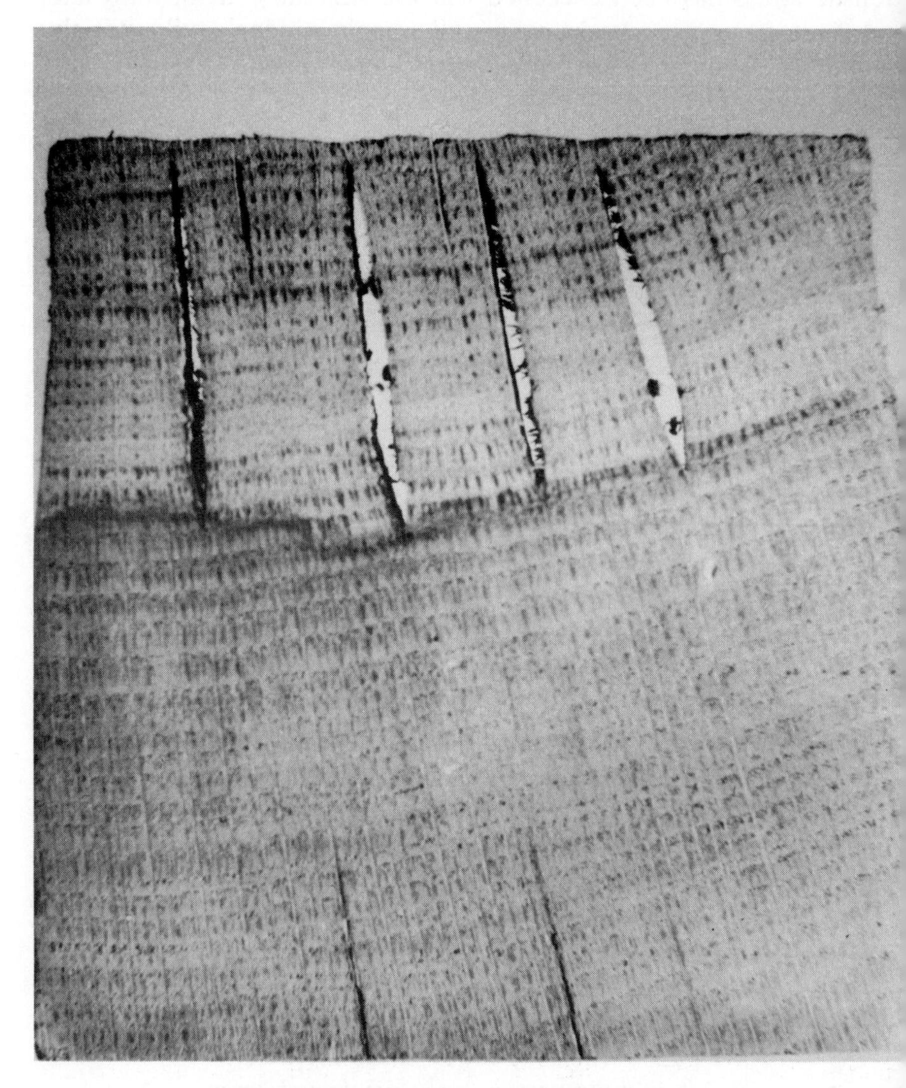

Fig. 96 Internal checking in oak. Note, pinching in of original surface checks. Honeycomb checks develop along rays.

of separation of certain of the growth rings. There are three basic causes for internal checking to develop: deepening of surface checks, extension of end checks, and the development of internal checks. In the case of softwoods which have been quarter-sawn the checking takes the form of failure at the juncture of latewood in one ring and earlywood in the adjacent ring.

Honeycomb checking does not occur frequently during air drying but when it does occur this is often due to extension of end and surface checking. The problem is that when drying stops the initial check may be very tightly closed and invisible to the casual observer but the wider, honeycomb form, hidden from view, may extend right throughout the length of the piece, rendering it useless.

Honeycombing is due to too fast a rate of drying, at the stage where the internal layers of the wood are shrinking in tension against already shrunk surface layers. See Figure 90(b). It should be noted that this form of honeycombing occurs below fibre saturation point.

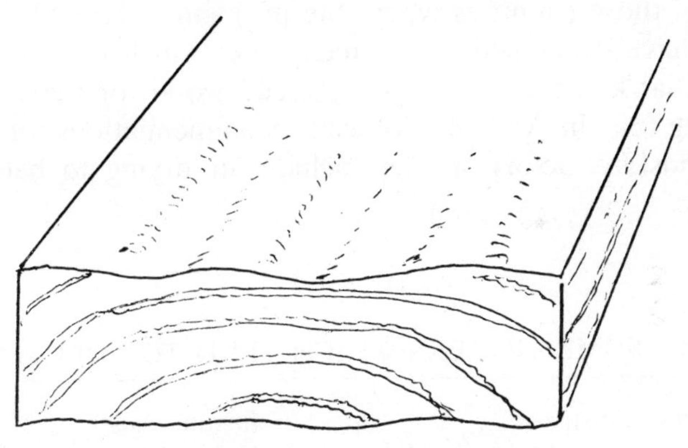

Fig. 97 Washboard effect in collapsed hardwood

Collapse

A serious fault that occurs during the kiln drying of certain woods from the green state is that known as collapse. Fortunately, we do not meet up with many timbers liable to this form of degrade but in North America the problem is a little more acute, and in Australia many indigenous species, particularly of the Eucalyptus genus are prone to collapse to the extent that seasoning methods have been specially formulated to control the problem.

The typical appearance of unplaned boards showing collapse is a corrugated or 'wash board' surface due to many of the internal cells collapsing. In this condition the wood is almost impossible to use since the cell walls are badly distorted or crushed. In the broadest sense, collapse causes abnormal shrinkage accompanied by distortion of the cell walls. In the early stages of drying, the free moisture leaves the cell cavities and as this moisture is evaporated from the surface of the wood tension forces are initiated which begin to exert a stress on the interior wood. If this stress becomes so great that it exceeds the ultimate strength of the wood, the cell walls are pulled inwards and collapse. It is this collapse of hundreds of minute cells which causes the typical wash boarding effect seen in collapsed wood.

Cell walls become plasticized when wet wood is heated; collapse is, therefore, more liable to occur when high temperatures are used in kiln drying; if air is present in the cells, collapse will not occur. Although for many years the phenomenon of collapse was not properly understood and much research was applied to the subject, presently, those countries where the problem is general with certain species, have developed drying techniques which basically allow for lower initial kiln temperatures for green wood for the first third of the drying run. In Australia, official recommendations for the drying of the most refractory species include air drying to below 30 per

TABLE 19: SOME TIMBERS WHICH TEND TO COLLAPSE

Ash, Alpine. Austr.	Jarrah. Austr.
Ash, Mountain. Austr.	Karri. Austr.
Black Bean. Austr.	Mahogany, Brush. Austr.
Black Butt. Austr.	Mahogany, Red. Austr
Brush Box. Austr.	Maple, Rose. Austr.
Cedar, Incense. Amer.	Oak, Eur. & Amer.
Cedar, Western red. Home & Amer.	Oak, Tasmanian
Elm. Eur.	Satinay. Austr.
Gum, Manna. Austr.	Sitka spruce. Home grown
Gum, Mountain. Austr.	Tuart. Austr.
Gum, Red. Austr.	Turpentine. Austr.
Gum, Rose. Austr.	Walnut. Eur. & Amer.
Gum, Spotted. Austr.	Wandoo. Austr.
Gum, Sydney Blue. Austr.	

cent mc prior to kiln drying with further recommendations for reconditioning.

Honeycombing may be associated with collapse where large aggregates of cells collapse and cause internal fractures. This type of honeycombing occurs above fibre saturation point and may normally be distinguished from the internal checking occurring below fibre saturation point, by the associated collapsed surfaces.

End Grain Protection

The protecting of the ends of individual boards and planks from splitting during air drying can be attempted in a number of ways, but the success of a given method will often depend upon the type of exposure rather than on the efficiency of the method. Wooden cleats for example will hold the end of a green board as if in a vice, but it does not follow that it will prevent the wood from splitting; on the contrary, it might encourage the wood to split simply because a wood cleat has its grain running at right angles to the grain of the wood; in other words, the wood can shrink in drying, the cleat cannot, so it holds the wood firm at the very end of the board but this shrinks from a point a little further along.

Metal or plastic cleats fixed to the ends of boards and planks tend to reduce splitting tendencies because they do offer restraint to shrinkage forces but buckle when the forces become too great.

Small round billets of wood can have their ends protected by tying tar paper or thick brown paper over them.

End Coatings

The most common form of end grain protection is by application of a water proof seal. For small quantities and small sections, dipping the ends of the pieces in melted paraffin wax is generally quite satisfactory, but for larger boards and planks the choice of coating while wet is not always satisfactory. The effectiveness of a seal depends not only on its ability to remain impervious to moisture but also on how many coats must be applied and the cost of so doing. Any seal to be effective must adhere to wet wood and possess a long service life without crazing or peeling and many types of normal wood finishes will fail under the conditions imposed on them.

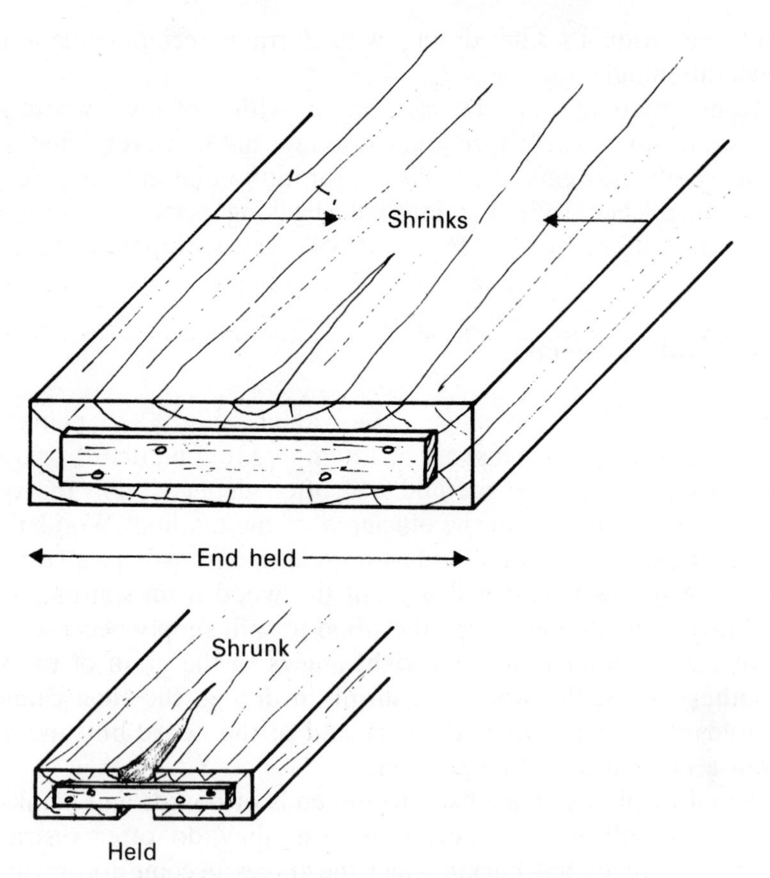

Fig. 98 Wooden cleats do not prevent end splitting. Cleat does not shrink longitudinally but holds wood firmly. Subsequent shrinkage away from board ends can encourge splitting.

The function of an end seal is to maintain a higher moisture content under the seal than would be the case if the board ends were left untreated.

It should be noted that end sealing compounds are applied to relatively green timber initially and thereafter the wood may remain exposed to varying atmospheric conditions in the open air for many months; also, much end coated wood eventually finds its way into a kiln. Accordingly, while wax based seals are often ideal for air drying, they can fail under the higher temperatures of a kiln by melting and seeping into the wood. On the other hand, quite a number of seals

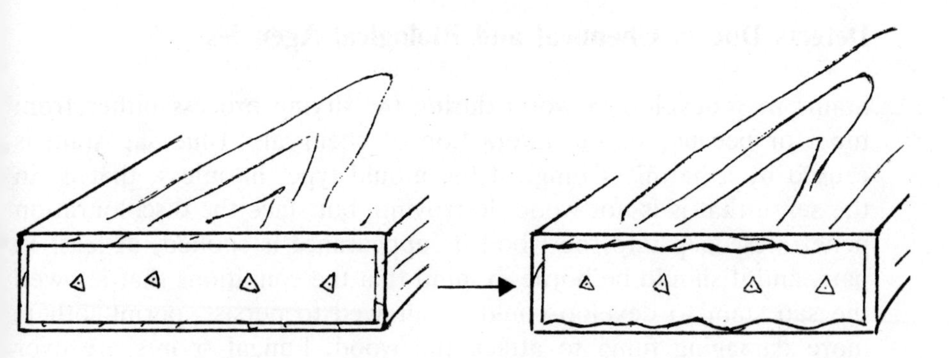

Fig. 99 A plastic or thin metal cleat restrains shrinkage and prevents excessive end checking. The cleat buckles as wood shrinks

appear to attract mould growth quite rapidly under yard conditions.

Paint systems can be effective water barriers provided a three coat method is applied i.e. one undercoat and two coats of gloss, but the cost must related to the value of the wood. Aluminium primer in a two brush coat application gives good results, but like some paints, blistering may occur under some conditions.

Bituminous compounds are generally very effective under all conditions but some types tend to remain softish, others tacky, and others give a hard coating. In laboratory tests carried out in the UK a product of the Mobil Oil Company called Mobilcer 'R' gave excellent results in terms of low cost and high effectiveness. The product is a thin hard micro-crystalline wax emulsion which dries into a white film when applied as a single brush coat. It behaved extremely well on test discs, but no field tests were made. A by-product of coal under the commercial name of Synthaprufe also gave very good results in these tests. This is a bituminous emulsion containing rubber which dries in a hard elastic film when applied in a single brush coat.

Generally speaking, end sealing is mainly used on the more exotic and valuable hardwoods of 2 inch (50mm) thickness and over; 1 inch (25mm) boards are less frequently sealed. Small logs or billets invariably require end grain protection and if these are sealed with a compound then usually the bituminous types are likely to stain the ends making trimming more costly; as a rule, the wax types are better in this respect.

Defects Due to Chemical and Biological Agencies

Stains may develop in wood during the drying process either from fungi or because of the interaction of chemicals. Blue sap stain is caused by a harmless fungi of the mould type; harmless, that is, in the sense that is is not wood destroying, but since the discolouration penetrates deep into the wood its appearance is spoiled, at least in part, and it should be borne in mind that the conditions that allowed the sap stain to develop could, if allowed to persist, permit other, more damaging fungi to attack the wood. Fungal spores are ever present in the air and often in the wood, and all they require for development is sluggish air, moisture, and food, and there is adequate food to be found in the starch content of sapwood and, in many cases in isolated patches in heartwood.

Sap stain develops rapidly, not only in softwoods but also in many hardwoods, especially light-coloured species, and although an attack may be arrested by drying it can be reactivated by rewetting. It is the practice today for many stain susceptible timbers to be treated with an inhibiting chemical prior to shipment, but this is not always an economic measure for the craftsman drying his own wood in the open air. Sap stain fungi will attack logs left lying around in warm weather and in freshly sawn wood, since the action of conversion exposes accessible areas of moist wood.

Since staining presents a greater risk during warm weather, it is best to attempt felling and conversion activities during late autumn and winter, but if a log has been produced in spring or summer than it ought to be converted as soon as possible. All saw dust should be swept from the boards and these should be put into stick immediately. If time is at a premium, then the wood should be end racked with at least 1″ (25mm) space between each board. In a good many instances in domestic situations, wood is frequently piled in areas near to lawns and plots of ground. A point that must be borne in mind is that garden sprinklers can often rewet an adjacent pile of wood and so create blue stain and water stain.

Square edged imported softwoods should not be close piled, especially the pines. There is often a tendency to do this, but these are very susceptible species and even under cover, say in a garage, if there is any sign of surface or other moisture on or in the wood, and the weather is warm, then sap stain can develop in a few days. Close piling of any wood, however, should be avoided if the moisture content is relatively high, because other forms of fungal based stain,

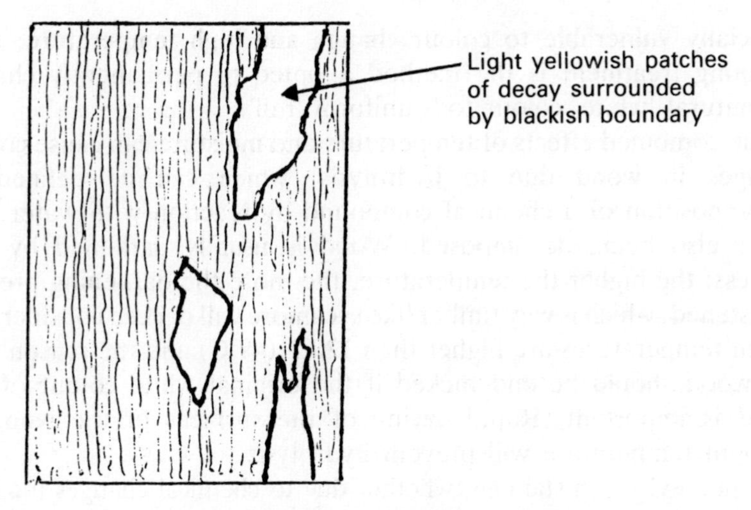

Fig. 100 The suitability of doty wood can be tested by inserting the point of a penknife in the discoloured areas. Brash fibres will break of short.

and worse, can develop. Decay is responsible for doty or spalted wood, and while in some forms it may represent a desirable feature to a wood turner or sculptor, it is less desirable to a cabinet maker and, if allowed to remain in the same, damp, closely piled state, the areas of yellow discoloured wood, originally hard and firm will become softened and pappy as the fungus takes over.

Some stains which develop during drying are thought to be caused by enzymes, particularly when temperatures are high as in kiln drying or relatively high in air drying. Sometimes the stain will develop under the outer skin of wood, in the slightly moister vapour zone and will only become noticeable when the wood is planed. In some instances, natural acids in the wood of a tree will promote staining; the pink colour sometimes found in freshly converted Douglas fir, is an example. In other instances, and oak is a good example, a yellowish or light brownish stain may be noticeable soon after converted wood has been exposed to air. Often, for no known reason, the stains will disappear after further exposure to light, but this cannot be relied upon since at other times the stain will be fixed. The same occasionally happens with beech, but this species is especially vulnerable to colour change and high temperature, high steaming treatment is the method adopted to permanently change the natural 'white' colour to a uniform 'red'.

The combined effects of temperature and moisture can cause colour changes in wood due to hydrolysis, which can be defined as decomposition of a chemical compound by reaction with water, the water also being decomposed. Wood is usually darkened by this process; the higher the temperature, the more the darkening process is hastened, which is why timber like sycamore will darken considerably if kiln temperatures are higher than 120°F (49°C) and the reason why the wood should be end racked if the original white colour of the wood is important. Rapid drying of the surfaces of the wood at ambient temperature will prevent hydrolysis.

Generally, much staining, whether due to chemical changes induced by Enzymes, bacteria, hydrolosis, etc, may be avoided, or reduced, by open stacking with sticks immediately after cutting. Most of these agents act in the presence of moisture and the more rapidly surface moisture is removed the better.

A characteristic of some hardwoods is the reaction of their colouring extractives to light, and especially to ultra violet light. While this reaction is not a defect of seasoning in its fullest sense it is considered appropriate to mention it since, on occasions, some colour changes give rise to complaint. Iroko, a West African hardwood is a case in point. When freshly sawn, its colour is light yellow, but on exposure it progressively darkens to a golden brown shade and tends to resemble teak. This comes about by photopathy, or response to light stimulus, and if a section of fresh sawn iroko has half of the piece masked by wrapping in brown paper, the exposed half will progressively darken and this will contrast sharply with the wood that is wrapped. If, after a day or two the mask is removed, it will be found there is a continuous lag between the two parts before colour becomes uniform and this can take a long time.

In the practical use of iroko, and especially where the finish is a natural one, if the wood represents a top or shelf on which ornaments or a clock are placed, it sometimes happens that the article is made and placed in service before the colour change of the wood is complete. Later, it is not unknown for a complaint to be made that when the clock or other is removed for dusting, the wood beneath is said to be 'fading'; the fact that the surrounding wood is imperceptibility darkening is seldom acknowledged. Iroko veneer can be affected in the same manner, because in store, in bundles, the lower leaves are invariably lighter in colour than the outer leaves. Teak and afrormosia behave similarly, but all woods darken more or less on exposure to light.

Water stain can spoil the appearance of timber drying in the open air; certain constituents found in wood and associated with staining are usually soluble and in many cases react with iron compounds to produce stains of varying colour. Timbers with a high tannin content, e.g. oak, sweet chestnut, western red cedar, yellow cedar, and others will react unfavourably to iron compounds released from rusty roof covers during rain, and black ink staining will result; and it is important to remember that iron contamination from polluted atmospheres can result from snow and wind blown rain gaining access to the wood within a stack.

Defects Due to Beetle Damage

Although not defects of drying, insect attacks may develop during air seasoning. Logs that are left lying around during warm weather, or converted timber stored in the open air, may suffer damage from beetle attack, but generally speaking, in the UK at least, beetle damage to stored wood is more usually superficial. There is an exception and that is in regard to the so-called Powder-post or *Lyctus* beetles. These can do serious damage to certain hardwoods held in stick in the open air or in open ended sheds.

In the UK the popular name Powder-post is restricted to two species of the *Lyctidae* family, *Lyctus brunneus* and *Lyctus linearis*, but a further group, the *Bostrychidae*, is recognised as the source of the *Bostrychid Lyctus* or tropical *Lyctus*. The two species of European *Lyctus* beetle breed in dead branches of suitable trees, old fencing, wood debris etc, its principal requirements being a wood species whose pores are at least 0.05mm in diameter so as to accommodate the ovipositor of the female insect during egg laying, and food in the way of starch. Accordingly, the sapwood of ash, elm, hickory, oak and walnut are ideal media to support breeding.

Attack occurs mostly in storage, in yards, sawmills, and other buildings with relatively open access during warm weather. Eggs are laid in suitable wood through end grain or by the female beetle biting a lead-in to the open pores on longitudinal surfaces. A typical life cycle in temperature climates is rather variable; in the open air or unheated storage shed it is usually from one to two years, but in heated premises the cycle may be considerably shortened, often to eight or nine months, the adult beetles emerging in March instead of the customary late May. On emergence, mating and egg laying continues the process; incubation under normal circumstances takes place in two or three weeks, with the last stage in the cycle, the pupal stage taking a further two to four weeks before metamorphosis occurs. The largest period of the life cycle therefore is larval stage during which damage to the wood occurs.

Lyctus does not attack green timber in the UK but infestation can occur once the wood begins to season. The damage caused is easily recognised; circular flight holes become noticeable in the sapwood about 1/16 inch (1.6mm) in diameter; fine yellow bore dust (frass) tends to collect on the wood which progressively becomes powdered, frequently between outer skirts of harder wood. If a little of the frass is rubbed between the thumb and forefinger it will have the feel of talc. Quite often, when the flight holes are first seen, it is concluded they are made by *Anobium* or Furniture Beetle, a much more serious pest in buildings. Examination of the frass as described is a simple way of distinguishing one from the other; in the case of *Anobium* the frass is much coarser and visual examination will reveal tiny lemon-shaped pellets present.

There are differences, of course, in the beetles, but where their grubs are similar in form, those of *Lyctus* will be seen to have a dark-coloured spot on either side of the tail segment. *Lyctus brunneus* and *L.linearis* are small reddish-brown to black-coloured beetles, often no more than 4mm in length but sometimes up to 7mm long.

Lyctus damage in stored timber is not regarded as seriously today in the UK as it was previously and it is the usual practice to apply remedial treatment if and when it happens, and this generally revolves around sterilising the wood in a kiln using a high temperature, high humidity treatment. In certain instances where *Lyctus* susceptible timber is to be used in special buildings, for example as replacement material in a building of historic interest, the sapwood is removed from the boards when the logs are converted and this is usually applied to oak. It is a point, however, the craftsmen should consider in relation to small amounts of wood he may be carrying which show obvious signs of infestation; the sapwood should be removed and burnt.

It has been mentioned that in this country, the term *Lyctus* is restricted to the families *Lyctidae* and *Bostrychidae*. In America the term is used more widely to describe many other types of wood

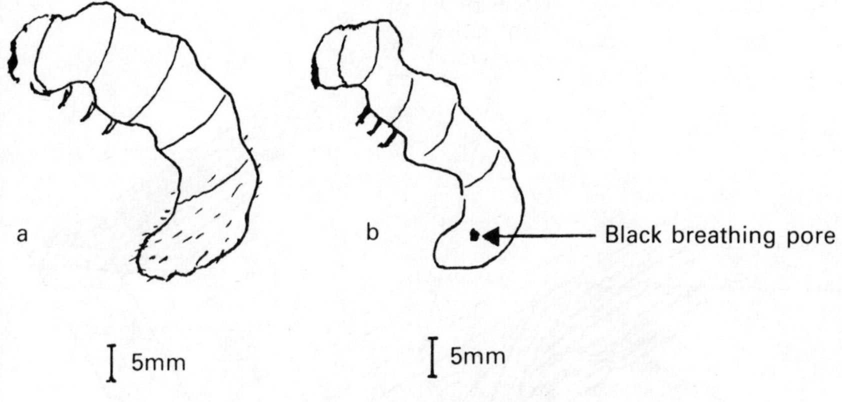

*Fig. 101 The larva of both furniture beetle (*Anobium*), (a) and* Lyctus *beetle (b) are similar. Both are curved, whitish, with three pairs of legs. They can be distinguished by fine hairs on* Anobium *and absence of black breathing pore, prominent on tail segment of* Lyctus. *Bore dust is like talcum powder in* Lyctus: *coarse, with lemon-shaped pellets in* Anobium

boring insects, some of which attack both hardwoods and softwoods and many will attack green timber. Because of this, official recommendations for kiln sterilising in the USA and Canada may differ slightly from those in the UK, but the high temperature, high humidity concept still applies.

An alternative treatment for timber in stick is to spray the wood with a suitable liquid insecticide. This is not always practicable for the small user whose stock is used up relatively quickly, say within eighteen months or so, because by the time the infestation is discovered and the treatment applied, the wood is probably wanted for use. This point is mentioned for a very good reason. In the merchanting of timber, a yard keeper may hold wood in stick for two or three years, sometimes longer; if signs of *Lyctus* appears he can either break down the stacks and have the wood sterilised or he can employ an annual spraying of the wood with a suitable insecticide. On the other hand, infestation may have occurred within a few months of the wood being sold and this only becomes obvious after the user has held the stock for a while. If the quantity of wood is

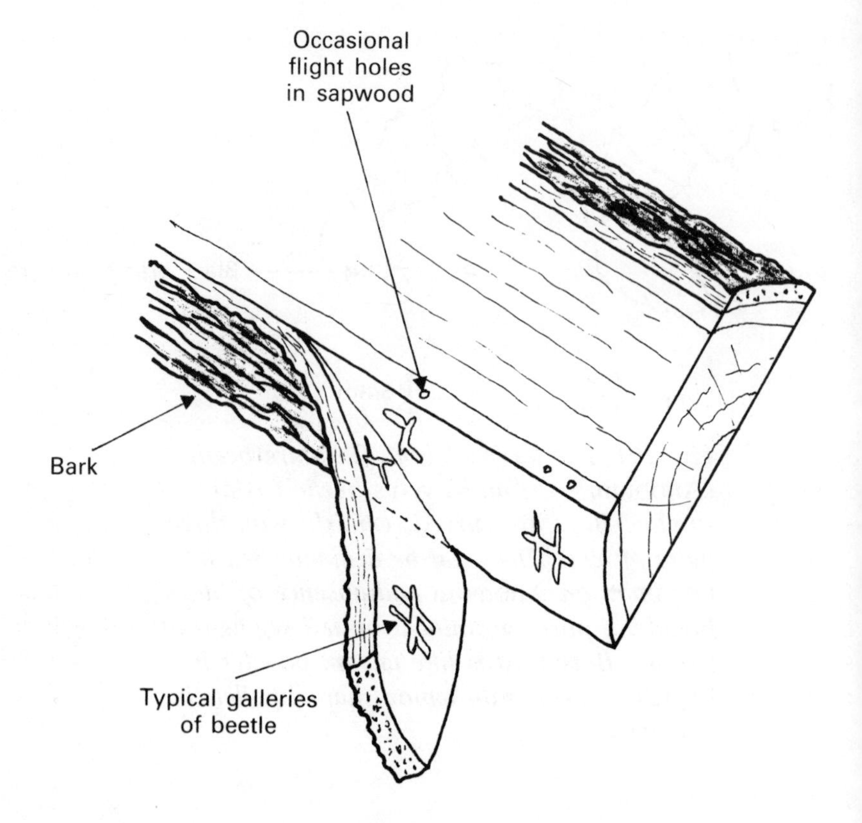

Fig. 102 Attack by bark borers is relatively light.

small, then the sapwood is best removed and burnt rather than to spend time and money on applying an insecticide, but larger quantities should be sent to a kiln for proper treatment.

The *Bostrychid* or tropical *Lyctus* are largely a problem for the timber importer in temperate countries when converting and seasoning logs. Unlike *Lyctus*, the *Bostrychid Lyctus* does not depend upon specially sized pores for egg laying and will attack all kinds of timber containing starch. The life cycle takes from one to three years to complete and while it is possible for a craftsman to buy wood that may contain a few grubs or eggs, the *Bostrychid Lyctus* does not, as a rule, breed in this country. Much tropical timber today intended for export in board form is dip treated in order to kill *Bostrychid*. For the user, this pest is generally of no real significance.

Treatment of Insect or Fungal Infested Timber

Wood which is affected by insect and/or fungal attack may be treated with suitable insecticides and fungicides. However, in many cases penetration of the wood is difficult and success is not always certain. Additionally, many chemical substances used for these purposes are expensive and experience has proved many unsafe for certain end uses.

A dilute solution of borax is generally effective both as an insecticide and fungicide. It is cheap, and as far as known, at the time of writing is safe to use, but is not permanent under damp or wet conditions.

An altogether safer and completely effective alternative is sterilisation within a kiln. This, of course, may not be a practicable suggestion to persons without kiln facilities. Nevertheless, sterilisation schedules are published by the Timber Research & Development Association, for those with access to suitable kilns or chambers. Generally, a few hours exposure at a temperature of 65°C at a relative humidity appropriate to the moisture content of the wood will be successful in eradicating attacks. Of course, the wood may be re-attacked if conditions again become suitable.

Bark and Debarking

The question of whether or not the bark should be removed from fresh sawn wood prior to air drying is debatable. If it is left on the wood it will be helpful in slowing down the drying and thereby preventing edge checking, and as the wood dries the cracking and peeling away of the bark from the wood will offer a guide to drying progress. On the other hand, bark left on may encourage attack from bark boring beetles, many of which cannot reproduce once bark is removed.

However, considering there are at least 20,000 species of one type alone, and bearing in mind that most of the damage bark borers cause in wood is restricted to the sapwood, it would seem that bark left on has no real significance in terms of potential beetle damage; in fact, in this country, the average woodworker, even in a lifetime spent at his craft, will have seen very little damage to his wood from this particular source. The question of whether or not to remove bark prior to seasoning can only be resolved when all the consequences are considered.

8
Other Methods of Seasoning

Although the traditional method of seasoning timber is either by drying it in the open air or by kiln drying at temperatures below boiling point, or by a combination of both, other methods have been developed, some proving to be uneconomical and failing to demonstrate commercial viability in terms of reasonable cost related to conventionally dried wood and, in some cases, reputed success has been restricted to the drying of easy drying species such as beech.

It has been explained in preceeding chapters that moisture is removed from wood in the form of vapour when exposed to air whose vapour pressure is lower than that of the wood and this moisture-moving force, depending on intensity sets up stresses which tend to distort, and in some cases to break down the wood. When other methods are devised to remove moisture from wood employing greater force it would seem the primary aim is to remove the moisture more rapidly without particular regard to the ability of wood to tolerate the conditions imposed, presumably on the assumption that all wood substance is the same. Basically, this is true, the substance of woody cell walls are all made up from the same material, but this is merely the substratum and the properties and complexities of wood inherent in every species, varies considerably with the thickness of the cell walls, the sizes of the cell wall pitting, the presence of tyloses and so on; in short, the relative permeability varies considerably from species to species, which will react differently, and sometimes violently, to forces which extend them to the proportional limit of their strength.

Science and technology do not stand still however, and new methods designed to dry timber economically and at a faster rate without undue loss of quality will doubtless be produced, but for the moment the alternatives to the conventional kiln drying method, except for dehumidifiers would seem of limited viability.

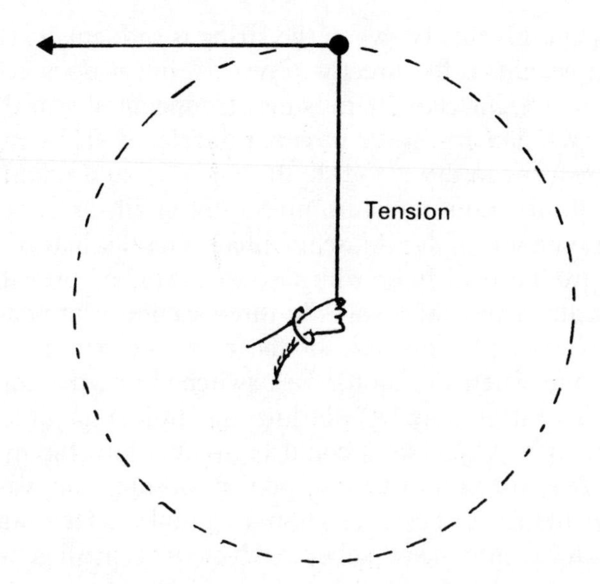

Fig. 103 A simple example of centrifugal force

Centrifugal Drying

In 1952 a report was published on the result of drying timber by what was termed, the Fessel spiral centrifuge, a German development which sought to employ centrifugal force as an adjunct to heat in order to remove moisture from wet wood. It was claimed that the equipment developed a radially and tangentially directed air stream which first removed the surface moisture from the wood, thereby creating a lower vapour pressure and thus encouraging the inner moisture to migrate rapidly to the surface, and, by centrifugal force, to be expelled from the wood.

From a purely physical angle the method was logical since the main use of centrifugal force is to separate liquids from solids, plus the fact that such a force would provide a variable air stream. Most people are aware of the basic principal of centrifugal force but it is worth mentioning the method by which the force is exerted. A simple example is that of a small weight rapidly rotated in a vertical circle at the end of a string held in the hand. The string is in a state of tension, and is thus exerting a force on the weight, tending to draw it towards the hand. This in-drawing force in theory was based on the supposition that the weight exercised an equal and opposite force outwards. On this theory, when the string is suddenly cut, one would expect the weight to fly directly outwards, but it does not do this in practice, the weight actually moving off tangentially in the direction in which it was moving at the moment of release. Thus, in employing this type of force to dry wood, both tangential and radial movement of air would draw out moisture uniformly in all directions.

The Eisenmann K.G. Maschinenbau Gesellschaft Company of Stuttgart produced a dryer which consisted of a vertical spindle to which was attached a steel cage, in three sections, into each of which timber was stacked. The base of the cage engaged a steel track in the floor, and when the motor was switched on, the cage revolved around the central spindle, whirling the timber at ever increasing speed. About 500 ft^3 ($14m^3$) could be loaded into the drying cages, and from the cost point of view, power consumption was said to be 400 W per hour, per 80 ft^3 (2.26 m^3) of timber. Heat and humidity were introduced into the chamber in which the centrifuge was situated, and it was claimed that for low operational cost, timber could be dried in a shorter space of time than by normal kiln drying, and splits and shakes would be avoided, with stresses reduced to a minimum. Further tests carried out in Prague, suggested that centrifugal drying could reduce the drying rate by 30 to 50 per cent of that normally taken by conventional kilns without undue degrade occurring. However, the tests were carried out in a laboratory in a small centrifuge, and the wood tested was beech and oak parquet flooring blocks.

Nothing further has been reported on the system in recent years and it is highly conjectural whether it would be both economical and efficient. Maintenance of the equipment would probably be costly because of the enormous forces imposed on the spindle and its bearings in the same way that intermittent breakdowns would offest savings in drying time. Provided there was adequate control of both heat and humidity, in theory the system simply increased air flow drying, but what has not been discovered, or at least not publicised, is the effect on the many hardwoods which do not release their moisture readily under conventional drying methods.

There are some wood turners who, probably on the assumption that domestic equipment spin-dries fabrics satisfactorily, occasionally leave a wettish section of wood to spin on a lathe during say, lunch break, and who contend this enables them to get on with the job much more quickly since the wood has dried in consequence. There is no doubt that free moisture would readily leave wood by this means, but really wet wood, above saturation point, ought not to be brought to a lathe for this purpose, firstly because it would take too long to bring the moisture down to a suitable level, and secondly, because of the unnecessary force imposed on the bearings. It is true that the action of turning a piece of wood does encourage it to lose a little moisture, but the wholesale adoption of this method to dry wood is not to be recommended.

Solvent Seasoning

Some years ago in America, experiments were carried out with the object of extracting oils and resin from knots in timber, so as to avoid the bleeding of resin through paint finishes. The solvent used was acetone, a colourless, limpid liquid, and quite accidentally, it was found that water was also removed from the wood at a remarkable rate. Further experiments were carried out on Ponderosa pine, (*Pinus ponderosa*) with very good results.

The pine, separated by specially fluted stick, was stacked on end in a tightly sealed extractor chamber and then continuously sprayed with hot acetone for several hours, after which the liquid was pumped out. Hot air was then circulated round the timber and residual solvent drawn off.

These experiments demonstrated that it is possible to dry this type of timber from 100 per cent mc or higher, down to 12 per cent mc as follows:

(a) for 1 inch (25mm) thick, spray for 24 hr, with hot air circulation for 4–6 hr.
(b) for 2 inch (50 mm) thick, spray for 36 hr, with hot air circulation for 6–8 hr.

While the process proved satisfactory with Ponderosa pine, there are one or two question marks related to the general drying of wood by this means. When the results of the tests were published, it was said that the quality of the wood and its colour were unaffected and the point was made that for subsequent preservative treatment of the wood, the permeability of the sapwood was improved, although the heartwood was not affected this way. Ponderosa pine happens to grow with quite a wide sapwood which might suggest that the method of drying would be much slower or perhaps ineffective if applied to less permeable woods. Nothing more has been reported on the subject in recent years.

Boiling in Oil

Another method of drying timber is by boiling it in oil, and perhaps worthy of mention. Heat is provided for the evaporation of moisture by submersion of the wood in hot oil. The operation involves the use of a steam retort equipped with an adequate heating device such as a steam coil and a vacuum pump to draw off the liquid. Various types of oily liquids have been used with boiling points above that of water. The wood is submerged in cold oil, and heat applied gradually until the oil reaches a temperature above 212°F (100°C) when the water in the wood is boiled off.

The treatment is of limited value since the dried wood is oily and unsuitable for most purposes. However, it is used to prepare transmission poles, railway sleepers, and similar for preservative treatment, the trade name being the Boulton process.

Super Heat

A variation on orthodox kiln drying is that known as drying in super heat, in other words, super heated steam is used instead of air to heat the wood and carry off the moisture. The system originated in America and was further improved and developed in Germany with varying success because, once again, while timber can be dried extremely rapidly by this method, other factors arise which largely militate against its wholesale commercial acceptance.

Radio-Frequency Drying

One of the most ambitious methods by which to dry timber, is the use of high frequency dielectric fields, which change electrical energy into heat inside of solids by rapid agitation of their polar molecules. The method differs in every respect from all other forms of wood drying. Traditionally, the surface of the wood must be heated to induce a withdrawal of moisture from the centre, but with RF drying the wood is heated uniformly through the thickness. With conventional drying, an enclosed chamber is used, with RF drying this is usually a continuous conveyance of the wood through a drying unit.

Basically, a RF drying installation consists of a short drying tunnel fitted with a high potential electrode above a conveyor belt with a condensor electrode below, the space being a dielectric field. The metal electrodes are fed from a generator which converts main power at 50 cycles per second to a million or more cycles, the high frequency oscillations heating the wood throughout and rapidly boiling off the contained moisture.

The system has been used in Switzerland to dry beech shoe lasts, and in America for drying turnery squares of birch and maple, but it would seem that the basic concept, while sound, had limitations, and further research was necessary. Accordingly, experimental work has, and is being carried out on modifications and improvement of the system.

Vacuum Drying

This method is based on the fact that the boiling point of water is substantially lowered when the pressure of the atmosphere over it is lowered; in other words, if the normal pressure of 14.7 psi (101kN/m^2) is lowered say to 2.9 psi (20 kN/m^2) the boiling point is lowered from 100°C to 60°C.

In early use of the method the wood was placed in a cylinder similar to the autoclave used for preservative treatments, the wood was heated up to 100°C by steaming or by immersion in boiling water. When the steam was shut off or the boiling water was drawn off a vacuum was applied to the wood and some of its moisture boiled off rapidly. The vacuum by itself does not cause drying; latent heat of evaporation is essential however, and as soon as the specific heat made available by the fall in temperature of the wood and the moisture in it is used up the drying ceases.

The method was limited because there appeared no easy way of conveying heat to timber in a cylinder under vacuum. In the absence of air, convection was almost absent and radiant heat around the perimeter only warmed up the outside of the stack. This was overcome originally by reheating the wood and drying was encouraged by applying a second vacuum. Drying to a final moisture content was achieved by repeated cycles which meant the times taken were not short enough to compensate for the small quantity of timber placed in the cylinder.

Research has continued in relation to vacuum drying and a number of companies are marketing a vacuum dryer in which heating is by means of aluminium plattens placed between each course of timber.

Fig. 104 Bollmann 9m^3 vacuum dryer. Photography, courtesy of Ludwig-Bollmann KG, West Germany

The air pressure in the sealed cyclinder is lowered and faster drying is said to be achieved at lower temperatures. It is claimed that the system is ideal for thick exotic hardwoods, especially oak, with kilning in some cases down to 20 per cent of the conventional drying times. A disadvantage is that variations in wood moisture content before drying produce varying levels of dryness in the final product.

Vapour Drying

A process known as vapour drying has been developed, and this enables wood to be dried by exposing it to the vapour of an organic chemical instead of air, maintained at high temperature within a tightly enclosed cylinder which forms the drying compartment. The principal chemicals used are xylene and perchlorethylene, the liquid of which has a boiling point above that of water. Their vapour provides the necessary heat for drying, and in circulating over the cooler wood condenses and liberates latent heat, the water being boiled off the wood.

The high temperatures obtained encourages very rapid drying of the wood; some ten to twenty times faster than orthodox kiln drying it is claimed. The process involves boiling the chemical in an evaporator from whence the vapours enter the drying cylinder. The moisture evaporated from the wood, together with any uncondensed vapour, is then passed though the separator tank containing a condenser which separates the water from the liquid drying agent. The chemical is then returned to a storage tank and is again available for circulation.

The idea was originally developed by the Taylor Colquitt Company in America as a pre-drying process to prepare transmission poles and railway ties for vacuum pressure preservative treatment, and from the green state the wood was dried down to 30 or 40 per cent moisture content in a very short time, something like 10 to 15 hours. From the drying cylinder the wood was placed directly into an autoclave to receive the preservative treatment. There were distinct advantages arising from the overall treatment; handling time was saved, and, whereas prior to the use of vapour, other forms of drying produced quite long and deep cracks in the wood, by the use of the chemical vapour it was found only fine checks developed.

Attempts have been made in America and Australia to modify and adapt the process to the drying of other types of timber but with only partial success. Permeable species have dried very rapidly, but impermeable species have proved difficult. In Australia, for example, attempts were made to dry collapse-susceptible eucalyptus, but it was found impossible to vapour dry these at elevated temperatures, but by reducing the wood temperature considerably, and applying a high vacuum, good results were obtained in about 30 per cent of the usual drying time.

The vapour used in the process is a gas, in the same way as air and hydrogen are gases, and the only advantage that one gas can have over another in relation to timber drying, is its specific heat and density, both at atmospheric pressure. The product of the two, i.e. specific heat times density, is its thermal capacity. Simple gases, like air, hydrogen, ammonia, etc, all have approximately the same thermal capacity per unit of volume when compared at the same temperature and pressure; therefore, the heating ability of all gases would be the same. Vapour from organic chemicals, on the other hand, has a much higher thermal capacity, sometimes three times higher. They are, however, very heavy and therefore a correspondingly greater horse power would be required to enable the fans to move the same volume. Thus, the higher power requirement could, in some circumstances, offset the gain in heating effect. A chemical like xylene, for instance, is over three and a half times as heavy as air, but its heat capacity is nearly three times that of air per unit of volume. High temperatures are necessary to keep organic chemicals in vapour form; in the case of xylene, at least 284°F at atmospheric pressure.

Chemical vapour drying would appear to be satisfactory, principally for reducing moisture content rapidly as part of a wood preservation system. For other end uses, especially with hardwoods, the need for conditioning at the end of a drying run would seem essential in order to release the wood from unwanted stress, and with the high power cost applicable to the method would tend to make the whole idea less profitable than normal kiln drying.

Chemical Seasoning

Selected chemicals can aid the drying of timber by retarding checking and splitting tendencies, and in some instances, can contribute to quicker kiln drying. Suitable chemicals are those which retain moisture on the surface of the wood, and also possess an anti-shrink ability. They should also be non-corrosive, and should not increase the electrical conductivity of wood, but unfortunately the corrosive effect of most suitable chemicals is fairly high.

Common salt, urea, invert sugar, ammonium sulphate, diethylene glycol, and others all have hygroscopic and anti-shrink properties, but common salt is probably the most effective although its corrosive effect on machines and tools when the wood is being worked is high, and a corrosion inhibitor really needs to be included in the salt.

Treatment is carried out on green timber, either by immersing the wood in a warm, saturated solution of salt in water, or by dry spreading the salt on the surfaces of the boards as they fall from the saw, close piling the wood allow the chemical to diffuse into the surfaces. Both methods require a day of two for the wood to absorb sufficient chemical.

The treated timber can be dried in a kiln or in the open air and checking is reduced because the vapour pressure at the surface of the wood is lowered, and while this encourages regular movement of moisture from the core of the wood, the surfaces are kept moist, and this, plus the anti-shrink effect of the chemical on the walls of the surface wood cells reduces tension stress and retards checking, especially in thick dimensions of Douglas fir and beech.

Treated timber can be dried more rapidly in the kiln because lower relative humidities can be employed without fear of encouraging checking, but drying time is usually extended when the treated wood is air dried.

The use of hygroscopic chemicals to aid the drying of difficult timbers was introduced some 40 years ago, but as structural adhesives have become more sophisticated so the use of laminating has increased and there is less need now for large sections of dry timber to be in one piece. There are exceptions, and the use of a chemical can often help provided the corrosion element is recognised. Originally, experiments were carried out on oak since the wood was expensive and liable to surface checking during seasoning. This use worked very well on thinnish stock, but on thick material the treatment tended to encourage the development of internal checks when the treatment was sustained, and this became necessary because of the difficulty in persuading the wood to accept sufficient chemical.

9
Stabilization of Wood

The basis of maintaining stability in wooden items installed under room conditions is that the moisture content of the wood at the time of installation should be the same as the average equilibrium moisture content of the environmental air conditions. If this is observed, then seasonal variations in the normal air conditions will not, as a rule, disturb the stability of the wood unduly. Once wood settles down to its environment in the average building, movement by shrinking and swelling is negligible, and long term, a degree of hysteresis develops. The term hysteresis is derived from the Greek, hysteren; to come later or behind, and in practice it means a retardation of effect when the forces on a body are changed. When wood has achieved an equilibrium with environmental air conditions in average rooms, any drastic change of these conditions will have little effect on wood short term because the wood will be slow to respond, and the hysteresis effect will be heightened by the finish applied to the wood; but long term, the wood could respond unfavourably if the environmental conditions were drastically altered, but this would be the fault of the air conditions, not of the wood if properly prepared in the first place.

The problem when working large sections of wood, mainly for turnery or carving, is the probability that it will develop checks and splits and since this type of wood is usually quite wet in the centre it is difficult to maintain a reasonable balance of stability throughout the working process. In turning bowls from wettish wood it is possible to reduce the volume of wood by rough turning well oversize so that there is less wood to dry, but the use of polyethylene glycol (PEG) can be an excellent medium with which to create a more stable material.

Polyethylene glycol is a wax type chemical, resembling paraffin wax and is related to permanent anti-freeze. It dissolves readily in warm water, is non-toxic, non-corrosive, and melts at 104°F (40°C); its fire point is high, at 580°F.

The use of polyethylene glycol on wood does not render it inert, but it can reduce shrinking and swelling tendencies to at least half the potential movement and often to less than the thermal expansion of most plastics, and it does this by bulking the fibres; furthermore, PEG suppresses decay where high concentrations are concerned, and has slight effect on a wood's physical properties and on gluing and finishing.

True stabilization of wood is achieved when it is soaked or impregnated with chemicals which replace the water held between the cellulose chains and then by heat and pressure the chemicals polymerise and the wood becomes altered and, in effect, becomes a plastic compound. This process is quite common with veneer but thicker wood requires more drastic treatment and this has been carried out by irradiation with gamma rays using polymers whose molecules become literally fused to the carbon molecules in the cellulose, but the wood undergoes severe changes in appearance and density but remains inert. Polyethylene glycol, when introduced into wet wood replaces the free water in the cell cavities and is then absorbed into the wood to replace the bound water in the cell walls, thereby tending to hold them relatively inert by the bulking action.

The first experiments on the use of PEG was carried out by Dr A.J. Stamm in America in 1934 but it was not until about 1959 that his test results were published. Stamm established that PEG in a molecular weight of 1000 gave the most satisfactory results and this is now normally used.

In 1946 the Swedish company Mo Och Domsjö– AB commenced research into the use of PEG to treat cellulose fibres and later they extended their work into a wide range of experiments involving plywood, chipboard and fibreboard. Little has been published in recent years on the results of the Swedish work, but PEG 1000 is being used commercially in America for stabilizing black walnut intended for rifle butts.

When PEG is introduced into a piece of wood it does not have a moisture retardant effect because as a chemical it is highly hygroscopic and actually attracts moisture. In a practical sense it replaces the water in green wood when this is at its greatest dimension but as the water is lost in drying the PEG remains, and under proper treatment conditions, eliminates something like 80 per cent of the normal dimensional change that could predictably occur under the most drastic end use conditions.

Polyethylene glycol 1000 is best used on green wood since this allows the maximum bulking effect to take place. In Dr Stamm's experiments it was found that the wood needed to take up about 35 per cent of its dry weight of chemical for optimum results, followed by storage under non-drying conditions to allow the chemical to equalise in concentration throughout the wood structure. However, for most practical purposes, the current recommendations are that PEG must be diffused into the wood in amounts of 25 to 30 per cent of the dry weight of the wood, an approximate concentration of the chemical being a 30 to 50 per cent (by weight) solution.

Wood that has been allowed to dry a little can be treated provided the wood is well wetted.

Dry or partially dry wood requires a period in which to take up sufficient moisture for it to be ready to accept suitable PEG treatment, but much depends upon the type of wood and its dimensions.

A relatively chunky piece of wood intended for turnery would not be suitable for PEG treatment since obviously it would be wasteful to treat wood that was mostly to be discarded; but if the wood was rough turned, and this applies especially to hollow ware, even if the wood was partially dry, much of the drier material would be discarded and the wetter wood left on would be more responsive to treatment. Where suitable material is available and is not wanted for a few weeks it should be stored in polythene bags to prevent loss of surface moisture.

Method of Treatment

1. PEG treatment should be undertaken as soon as possible after conversion from the log, but some pre-cutting of the sections to oversize dimensions is desirable. As a guide, turnery sections should be rough turned, allowing about 1 inch (25mm) oversize for solid pieces and a similar amount for a ½ inch (12.5mm) finish for hollow ware. For carvings, about ¾ inch (18mm) should be allowed over the largest dimension.
2. Wood is treated by soaking in a container in PEG solution. The container must not be metal, except certain kinds of stainless steel, because certain constituents in wood are liable to react unfavourably with metal, chemical and water, and thus create staining of the wood. Plastic, fibreglass, glass, and earthware can be used as a vessel, even a plastic dust bin, or a wooden vat can be made and lined with plastic

sheeting. The vessel should be fitted with a lid to reduce evaporation tendencies.
3. Make up a 30 or 50 per cent (by weight) solution of polyethylene glycol 1000 with warm water using a vessel large enought to take the envisaged number of pieces. These must be fully immersed and if necessary be weighted down with rocks or flints. The two common solution strengths have specific gravities of 1.05 and 1.093 respectively and where fairly large quantities of wood are under treatment it is advisable to check the s.g. by means of a hydrometer, topping up with water and PEG when necessary.
4. The time allowed for treatment will depend upon the type of wood, the temperature of the solution, and the dimensions of the wood. For large scale PEG treatment a vat fitted with an electric device to heat the solution, a built-in thermostat, and a small pump to circulate the solution periodically will give the best results but this is not essential for treating a few pieces at a time. Room temperature can be quite satisfactory, but standing a small vessel such as a plastic bucket in a warm airing cupboard will shorten the soaking time, while hand stirring of the solution is quite effective if carried out regularly.

Soaking Period

Although in simple terms PEG bulks the fibres of wood, different species of wood, quite apart from their dimensions, respond differently to the passage of liquids in or out of their bulk. This is related to their relative permeability, an aspect which has a direct bearing on the ease or otherwise with which moisture is drawn from wood during the seasoning process. PEG treatment is operating under atmospheric pressure and the resistance of some woods could be considerable. In tests carried out at Forest Products Laboratory, Madison, Wisconsin, on the uses of PEG the following treatment schedule was suggested for black walnut to give the desired 25 to 30 per cent concentration of PEG in the finished product.

It should be noted that the above suggestions refer to discs (i.e. round segments of wood in which the broad faces are end grain surfaces). Diffusion of PEG solution in the direction of the wood fibres is nine to fifteen times faster than it is across the fibres, so a wood disc will take up the solution much faster than would be the case with a longer section with a preponderance of side grain. Taking this into account and considering the permeability of wood and its amenability to accepting water-borne preservatives, the following guide is offered as to suitable soaking times for various wood species.

The above recommendations are given purely for guidance. The relative permeability of different timbers refers more directly to side grain which means that the soaking times given above are based on the assumption that a higher proportion of side grain will exist in the average sections likely to be treated. Some experimentation in soak periods is therefore necessary in the absence of proper data, other than for discs, being presently available. An adjustment would be needed for example if red oak in disc form was a subject for treatment since while it is extremely resistant to impregnation through side grain, its absence of tyloses in the vessels makes its end grain relatively porous.

TABLE 20: SUGGESTED SOAK PERIOD FOR WALNUT DISCS

Concentration	**Temp.**	**Up to 9″ (229mm) diam. and 1″ – 1½″ (25 – 37mm) thick**	**More than 9″ (229mm) diam. 2″ – 3″ (50 – 75mm) thick**
		Days	Days
30%	70°F	20	60
50%	70°F	15	45
30%	140°F	7	30
50%	140°F	3	14

Temperatures have a definite bearing on the ability of the wood to accept PEG; the warmer the solution the more rapid the uptake of PEG, so any action that will raise the temperature above ambient is important. Placing say a bucket of material in the solution on or near a heating source in winter, or outside in strong sunshine will reduce soaking time. If by increasing the temperature this allows some evaporation to occur, then the solution must be topped up and the use of a hydrometer is obviously helpful. It is helpful too, to keep to hand an untreated control sample of the same material and dimensions with which to compare results. This will not only provide a guide as to the efficiency of the treatment but may indicate possible adjustment needed to soaking time.

After treatment the wood should be allowed to drain prior to drying. Since the treatment is designed to hold wood against shrinkage

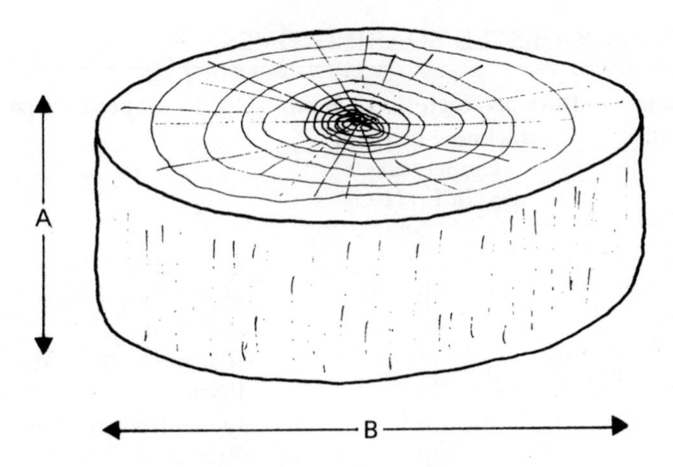

Fig. 105 Diffusion of PEG is 9 to 15 times faster in the fibre direction A, than across the fibres, B. Thus a wood disc with a high proportion of end grain will take up the solution more readily than a long grain section.

it can be dried much more drastically than would normally be the case. Depending on dimensions, drying may take from a fortnight to a month or so, but in a good many cases, once the surface of the wood has dried off it can be worked, letting the remainder of the bulk equalise in its own time. Properly treated sections can be force dried by placing them on, or very close to a suitable heat source, say a radiator or electric fan heater, without danger of splitting or checking, but an enclosed oven should be avoided since there could be a darkening of colour of the wood. The solution that is drained off after treatment can be re-used, but it should be stored in either a glass or plastic container with a sealed lid.

PEG treatment will not spoil the natural colour of wood, nor can it be over-treated. Tools may pick up some waxy deposits but since PEG is soluble in water these may be quickly removed by washing in warm water. When carving PEG treated wood, sponging the wood surface with warm water every so often will reduce or eliminate any tendency for the tools to gather wax on their cutting edges. Coarse wet and dry sandpaper used in turnery may tend to collect wax particles within the grit. Again, these can be removed under running warm water using a brush to aid the process.

TABLE 21: SUGGESTED SOAKING TIMES

* Solution strength	Days per each $\frac{1}{2}''$ (12.5mm) thickness at room temperature say 70°F (21°C)	Species type
30%	7	Permeable
	10	Moderately resistant
	12	Resistant
50%	15	Extremely resistant
	5	Permeable
	8	Moderately resistant
	10	Resistant
	12	Extremely resistant

* 4.5 lbs of PEG per gallon of warm water will give a 30 per cent solution. 10lbs per gallon will give a 50 per cent solution, or 450 grams per litre and 1000 grams per litre respectively.

PEG treated wood can be glued once it is dry enough, but some adhesives are a little unreliable. For instance, the most popular adhesives used by the craftsman today are probably the white, polyvinyl-based type, and these do not perform well on PEG treated wood; while the urea-formaldehyde adhesives bond well initially, they are not resistant to water and since PEG attracts moisture, a UF glue line would tend to break down under humid conditions that could apply in say, a kitchen or bathroom, whilst in other conditions of use a urea adhesive would be satisfactory. The most satisfactory glues, however, are reported to be the two-component resorcinols and epoxies.

Glue works well on PEG treated wood only if the contact areas are properly prepared by degreasing with a solvent to cut the wax and raise the fibres. White spirit could be used as it works very well in degreasing teak prior to gluing, and would be satisfactory with PEG, followed by a light sanding.

PEG treated articles can be finished with conventional polyurethane varnishes or modified polyurethanes, applying thin coats with a light sanding between coats, while moisture curing polyurethane varnish is recommended where the finished article is likely to be exposed to high humidity. It is reported that of all the oil finishes tested at Madison, only a Danish oil penetrating finish worked satisfactorily over PEG-treated wood. Beeswax and carnauba wax polish can be used satisfactorily after the wood has been degreased with white spirit.

PEG 1000 can be used for a variety of purposes besides turnery and carving. The original work carried out by Stamm was aimed primarily at increasing the stability of walnut rifle stocks used in competition shooting, since even the slightest movement of the stock could place the barrel out of alignment. This treatment has proved highly successful and it could be extended to other sporting goods such as laminated racket frames for tennis and for racket presses, bows for competition shooting, and so on; but since PEG is water soluble and could be leached out if wetted badly, naturally, the finish used on such items would need to be water resistant at least. In interior situations where there might be periodic changes of atmosphere, wooden items like shoe lasts, or wood moulds could be made from PEG treated wood, and it is probable that such use could be of advantage in pottery works to improve the stability of wood racks.

PEG can be obtained in small quantities from specialist wood turnery suppliers.

10
Protection, Care and Use of Dry Wood

There is immense satisfaction to be gained from producing precision woodwork but good craftsmanship implies that not only is the finished article good for its intended purpose, but it will remain so in service.

It is not the easiest task to always achieve precision in stability for example, because service conditions may be beyond the control of the maker of the goods and this applies also to storage conditions where the woodwork may be held for a time before delivery to a customer. It is always a good idea to try to work to a specification satisfactory to both parties, since if this is adhered to through all the manufacturing and finishing processes it does form a sound basis to support or contest a dispute. *British Standard BS 1186: Parts 1 and 2: Quality of Timber and Workmanship in Joinery* is a good example, but there are several others.

The moisture content of wood must be at a satisfactory level at all times which goes without saying, but the point that is either overlooked or is not properly understood, is the effect that atmospheric changes have on wood; everyone knows that warm dry air will encourage wood to shrink and cool moist air to make it swell, but the prime factors of temperature and relative humidity are often misunderstood or are disregarded because variations that are suspected in air conditions appear to be beyond control and therefore must be tolerated. To some extent, especially in buildings, this is true; a ceiling in a home, or an entrance door, will more often than not have different atmospheres on either side likely to create bowing tendencies in the material, but the point is, that if we appreciate that there are some situations more critical to the use of wood and wood based materials than others we can try to eliminate these and contain the others.

Wood deemed to be dry enough for different forms of accepted use generally falls into a moisture content range of about 15 per cent for air-dried stock and around 10 per cent for stock dried in a kiln. The service equilibrium mc will differ from these values according to the situation, so it is obviously important to know how much moisture is in the wood when it is taken into the workshop so as to take suitable steps either to allow the wood to dry a little more, or to see to it that its initial moisture content in maintained after receipt.

Outdoor Storage

Although air-drying represents outdoor storage and in this sense the wood ought to be allowed to dry as uniformly as possible without disturbance, there comes a time when it is considered dry enough to be removed from stick and to be piled elsewhere in a close-piled manner. Some material might of course be taken directly from stick and loaded on a vehicle for delivery to a customer, or it might be built up into conveniently-sized packages for eventual loading with a fork lift truck. As a temporary measure it could be wrapped in a water-proof paper shroud, but if the wood is close-piled and left either in the open air or in say, a relatively open car port, as could be the case with fairly small quantities, for lengthy storage periods, shrouding of the wood with a sheet material, especially if this was polythene could defeat the object of protection since in warm weather the wood would tend to sweat and so encourage moulds to form on the wood, and, in some cases lead to sap stain in the sapwood portions of the boards. Where wood is stored close piled in the open it should be a short term expedient with only a top cover.

On building sites, some timber and joinery inevitably has to be stored in the open for a time; in fact it is often difficult to assess whether it is better to leave the material in the open or to store it temporarily in the building once this has the roof on, or in the favourite place, in the garage. With wet-constructed buildings, the relative humidity in a brick-built house or garage could be very high during the summer months, even when doors and windows have yet to be installed and any wood goods would tend to pick up moisture if stored therein. Strictly speaking, deliveries of interior joinery and fitments should be phased in and only accepted after the heating has been turned on and the fabric has had a chance to dry out. Where wood in any form has to be stored outside, then first of all it must be placed on suitable bearers to keep it off the ground and then be adequately covered, with suitable ventilation, under the polythene which is generally used to prevent condensation. Proper measures must also be taken to store critical joinery items in such a way that they do not distort of suffer damage.

Interior Storage

When air-dried timber is removed from stick and taken into covered storage this is done for one of two reasons; either it is held as stock for sale to a customer or it is held against fairly immediate or ultimate use. Where small quantities of air-dried wood is brought inside, as is the case with a craftsman or similar user, this is usually done in dry, warm weather, and the material is often placed in a shed, garage, or workshop. Under these conditions, protection from the elements is

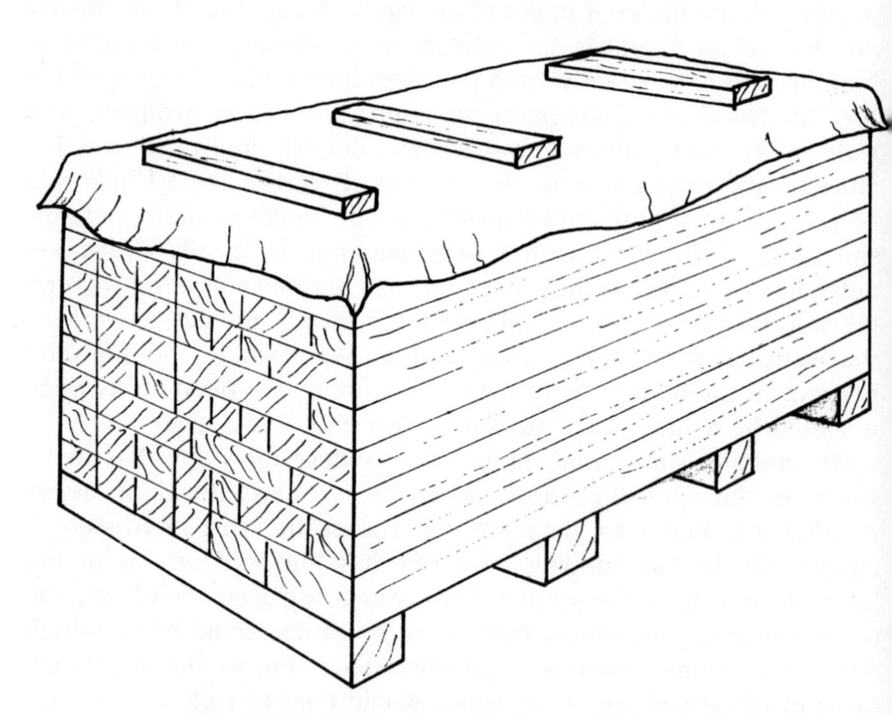

Fig. 106 Air-Dry wood, if stored close-piled in the open, should not be completely shrouded with polythene. As a short-term measure the top only should be covered to avoid sweating. There should be battens between the cover and the wood.

unnecessary, but due regard must be paid to the air conditions of the storage since further drying of the wood will now be encouraged to take place.

The air conditions in an enclosed space are governed by several variable and fluctuating factors. These include the temperature and relative humidity of the incoming air together with daily fluctuations, the effect the entering air will have on the internal air, and the number of air changes due to ventilation.

As an example, assume a wooden shed or workshop is to recieve some air-dried wood with an average moisture content of 15 per cent. The outside air has a temperature of 70°F (21°C) and a relative humidity of 70 per cent and when this air is drawn into the shed or workshop it is heated to 80°F (27°C). It will be noted by reference to Table 22 that if no moisture is added, the internal air now has a relative humidity of 51.1 per cent, in other words, the air is now drier and can therefore hold more moisture. Further reference to Table 23 will indicate that 51.1 per cent RH corresponds to an approximate equilibrium moisture content value for wood of 9.5 per cent, and it is to that value the wood will now adjust; the time in which it does so depending upon how rapidly, or slowly, air changes between external and internal air takes place.

The above example would represent quite good conditions in which to handle air-dried wood that needed to be conditioned further, say for furniture, but more often than not, the same air entering the type of workshop or storage area described, could be heated to 90°F (32.5°C) and this would mean the internal air RH would be lower still, i.e. at 37.8 per cent, equalling an equilibrium mc for the wood of rather less than 8 per cent. This would tend to dry the wood quicker, but it must be remembered that the temperature mentioned corresponds to the better months of the year, and one of the problems in storing wood is to get the balance of moisture right in the storage area at all times of year.

Suppose in the above examples, it was winter time, and the air entering the storage area was at 40°F (4.5°C) with a relative humidity of 80 per cent and it was now artificially heated to 70°F (21°C). Until more moisture was added, the RH would fall to 28.4 per cent corresponding to an EMC for wood of about 6 per cent which would make for very rapid drying conditions which would be controlled entirely by ventilation leading to air changes. If windows and doors were kept closed as much as possible to conserve heat, then the RH would increase and the wood dry slower.

TABLE 22: EFFECT ON RELATIVE HUMIDITY, OF HEATING AIR TO VARIOUS TEMPERATURES*

Incoming Air		**Heated to**				
Temperature	**Relative Humidity**	°F 50 °C 10	60 15.5	70 21	80 27	90 32.5
		Relative Humidity				
	%	%	%	%	%	%
40°F 4.5°C	30	20.9	14.8	10.6	7.8	5.7
	40	27.9	19.8	14.2	10.4	7.7
	50	34.9	24.7	17.7	13.0	9.6
	60	41.8	29.6	21.3	15.5	11.5
	70	48.8	34.6	24.8	18.1	13.4
	80	55.8	39.5	28.4	20.7	15.3
	90	62.7	44.4	31.9	23.3	17.2
50°F 10°C	30		21.3	15.3	11.1	8.2
	40		28.3	20.3	14.9	11.0
	50		35.4	25.4	18.6	13.7
	60		42.5	30.5	22.3	16.5
	70		49.6	35.6	26.0	19.2
	80		56.7	40.7	29.7	22.0
	90		63.8	45.8	33.4	24.7
60°F 15.5°C	30			21.5	15.7	11.6
	40			28.7	21.0	15.5
	50			35.9	26.2	19.4
	60			43.1	31.5	23.3
	70			50.3	36.7	27.1
	80			57.4	42.0	31.0
	90			64.6	47.2	34.9
70°F 21°C	30				21.9	16.2
	40				29.2	21.6
	50				36.5	27.0
	60				43.8	32.4
	70				51.1	37.8
	80				58.4	43.2
	90				65.7	48.6

* Note consistent decrease in relative humidity as temperature is increased.

Air conditions suitable for the well-being of wood, either as a material or as a finished product, involve adequate control of temperature, relative humidity and ventilation, and the key factor is relative humidity. In all types of indoor situations there are a number of air changes during each 24 hour period and, as already mentioned, as cool air enters the building its moisture holding capacity is increased as it is warmed to room temperature, and if no moisture is added the air becomes very dry. Some moisture is always available, even from human sources, since we all breathe and perspire; but in most interior locations other activities such as cooking and washing release moisture into the air, as do some forms of heating. To cater for excessive humidity we rely upon ventilation to dispose of a proportion of moisture laden air, but if we ignore the signs then condensation and sticking doors and drawers problems arise.

It is a simple matter to measure temperature with a thermometer, and just as easy to measure relative humidity by means of one or more, inexpensive direct reading dial-type humidity meters, but it is often not so easy to regulate temperature and ventilation, which control relative humidity. In buildings with air-conditioning the air is controlled automatically to comfort level where, irrespective of temperature, the relative humidity is usually about 50 per cent, a level which maintains wood around the 9 to 10 per cent mc mark, and this is the level which should be sought in homes and workshops, by attempting to control the factors which alter relative humidity drastically.

TABLE 23: EQUILIBRIUM MOISTURE CONTENT RELATED TO RELATIVE HUMIDITY IN TABLE 28*

Relative Humidity %	**Equilibrium Moisture Content %**
5 to 14	1.3 to 3.3
15 to 24	3.7 to 5.5
25 to 34	5.8 to 6.8
35 to 44	7.0 to 8.3
45 to 54	8.5 to 10.0
55 to 66	10.1 to 12.0

* The above values represent interior air conditions when exterior air at a given temperature and relative humidity enters a building or other enclosed area and is heated with no further moisture added. It further demonstrates the drying power of different air/vapour mixes.

Fig. 107 Direct reading dial humidity meter. Photograph, courtesy S. Brannan & Sons Ltd

Air Conditions Applicable to Various Industries

When wooden items are demanded by a customer quite obviously the condition of the finished product must be compatible with the service conditions and, as already mentioned, the work ought to be carried out in conformity with a suitable specification. In a surprisingly large number of cases, standard specifications covering wooden items appear to stress moisture content in average terms, that is to say at a level that generally pertains in given areas of a building, so for interior situations, recommendations for satisfactory moisture content is generally given as 9 to 11 per cent for furniture and 10 to 12 per cent for joinery. These values are reasonable in the sense that if the specification is followed in detail then the finished item should remain serviceable. The problem, however, is that the emphasis on controlled air conditions in buildings has increased in step with higher standards of insulation against heat losses; heating and ventilating engineers have been obliged to recommend air conditions that are satisfactory to a product or use and the fact that wooden items may have to be placed or fitted in a particular environment becomes secondary to the main purpose of the building.

There is a point here that must be recognised, and that is, in specifications pertaining to wood, the criterion on which the suitability of the finished product is judged is moisture content of the material, whereas the criteria on which an engineer assesses the suitability of a building is relative humidity. Temperature in both cases being considered simply a matter of comfort to the occupants of the building. The factors of temperature and relative humidity correspond of course to equilibrium moisture contents of wood and other organic materials and if this is related to a given use it will be seen that a wooden item designed for general use could be out of step for specialised use.

Suppose a simple item such as a cupboard is made. If it was then placed in a home with air conditioning set at the usual level of 50 to 60 per cent RH the equilibrium moisture content (EMC) for wood in these conditions would be from 9 to 11 per cent, so if the item had this sort of value when installed it could be expected to function as intended. On the other hand, that same cupboard might be required in a meat processing plant or in a factory making cardboard cartons. In this latter situation, if the plant was air conditioned, the cupboard could be expected to perform satisfactorily because for that industry and the materials it used, the RH would be similar to a domestic dwelling. Not so in the meat processing activity, because here the air needs to be moist in order to maintain the quality of the product, and the same cupboard could suffer in consequence since the EMC for the wood would now be at least 18 per cent and maybe more, Table 25.

It is the practice today for more and more public and industrial buildings to be fitted with humidifiers in order to provide a proper balance of moisture in the air and woodworkers should be aware of this and ascertain what the end use conditions are likely to be for any item before going into production. Table 24 offers a guide to likely levels of relative humidity applicable to different situations, while Table 25 gives the corresponding equilibrium moisture contents for wood exposed to relative humidity levels.

There are occasions where a craftsman delivers a few special items of turnery to a shop or stores. In the smaller type retail outlet, these items might be displayed in a window fitted with a heating device to prevent condensation. This does not help the wood, and it is often impossible for the craftsman supplier to know what is happening to his work if the shop is in another district. It usually becomes a matter

TABLE 24: LEVELS OF RELATIVE HUMIDITY APPLICABLE TO VARIOUS INDUSTRIES AND PRODUCTS

	%		%
Air Conditioning	50 – 60	Hospital Wards	50 – 60
Animal Rearing	45 – 60	Hospital Operating Theatres	30 – 50
Antiques	50 – 55	Knitware	50 – 60
Apple Storage	85 – 90	Lace	50 – 60
Bakeries	60 – 80	Leather	55 – 65
Bowling Alleys	45 – 55	Letter Press	45 – 55
Breweries	65 – 75	Lithography	50 – 60
Carpets	50 – 60	Meats	90 – 95
Cartons	45 – 65	Museums	50 – 55
Ceramics	45 – 50	Photography	45 – 55
Cigarettes	50 – 60	Printing	50 – 60
Cigars	65 – 70	Radium	40 – 50
Egg Storage	70 – 90	Rayon	45 – 55
Electronic Computers	45 – 55	Silks	50 – 60
Film Processing	50 – 60	Tea	55 – 65
Film Storage	45 – 55	Textiles	50 – 65
Florists	50 – 65	Tobacco	55 – 65
Food Storage	60 – 95	Wool	55 – 65
Fruit Storage	70 – 95		

TABLE 25: EQUILIBRIUM MOISTURE CONTENT FOR WOOD RELATED TO INDOOR AIR CONDITIONS

Relative humidity %		**Temperature**		
	°F	40	to	90
	°C	4.5	to	32
			EMC	
90		18.0	to	21.8
80		17.0	to	16.0
70		14.0	to	13.0
60		11.2	to	10.5
50		9.3	to	9.0
40		8.0	to	7.4
30		6.3	to	7.3

of tact and diplomacy if a complaint is made, but it is a point that ought to be resolved at the time of delivery, that the quality as delivered should be preserved by proper storage and display; if there is a subsequent complaint at least the supplier has some safeguard.

Moisture Content of Wood in New Houses

Legislation exists in the form of Building Regulations, Codes of Practice and so on, in order to ensure that materials used in building perform the function for which they were intended without premature deterioration. Recommendations are made as to the moisture content of structural timbers and joinery which should not exceed a particular level at the time of erection or installation, and these assume a degree of movement, especially by shrinkage, tolerable after erection.

These recommendations and requirements work very well if adhered to because the initial woodwork that is installed in a home such as roof and floor structures and staircases are restrained against excessive shrinkage by virtue of their construction. Problems of unwanted shrinkage are more likely to occur after the building has been handed over and occupancy begins, since at this stage, and especially during autumn and winter months, heating of the premises is often increased beyond the normal for a lived-in building, while ventilation is also increased to assist the drying out.

It is generally easier to appreciate temperature variations than it is to understand the complexities of relative humidity fluctuations; everyone can reasonably assess what is meant by cold, warm or hot, since these are conditions to which the human body responds in terms of comfort, but on the other hand, we are generally insensible to relative humidity. When this is low, it is usually considered an obscure reflection on excessive heat, and if it is high, it can pass unnoticed, except by persons afflicted by chest complaints. The point is however, that heating up the air of rooms considered damp, while essential, should be kept to reasonable limits and taken slowly.

It has already been stressed that if fairly dry air is drawn into a room and is then heated to a higher temperature its moisture holding capacity is increased, but if ventilation is likewise increased, some wood in the building could react unfavourably, as could plaster work.

Table 26 gives official recommendations for levels of moisture content for furniture and joinery at the time of installation. The last column has been added to indicate the relative humidity levels which are needed to maintain the timber at the recommended level.

If thought is given to the various values it will be seen that if by increasing the heating to dry out the building this results in an increase in RH then firstly all the woodwork will tend to swell, and secondly, will tend later to shrink back to somewhere around its original condition. This in itself is a basic requirement but it is the reaction of the wood to the increase in RH that may create problems. Take the example of upper floor joists. These are invariably set on edge so that the widest face with the most movement potential is vertically opposed to the floor above and the ceiling below. It will also be noted that the permitted moisture content of the ground floor joists in suspended flooring is three points higher that those for the upper floor, while a difference of only 5 points separates the relative humidity levels. Suppose, in drying out the building the relative

TABLE 26: RECOMMENDED MOISTURE CONTENT LEVELS FOR WOOD INSTALLED INDOORS

Application	Moisture Content %	RH equalling MC %
(a) Flooring		
Intermittent heating	12 – 15	65 – 75
Continuous heating	9 – 12	48 – 65
Under-floor heating	6 – 10	30 – 50
(b) Furniture, Cabinets etc.		
Intermittent heating	10 – 12	50 – 65
Continuous heating	9 – 11	48 – 60
(c) Joinery, (All, including doors)		
Intermittent heating	12 – 15	65 – 75
Continuous heating	10 – 12	50 – 65
(d) Carcassing and framing*		
Roof joists & rafters etc.	15	75
Upper floor joists		
Ground floor joists	18	80

* These recommendations also cover wooden extensions built by the house holder and include framing and sheathing, roof boarding and so on. After completion and drying out, all other woodwork will fall into categories, a, b, and c.

humidity is increased to 85 per cent due to inadequate ventilation, then the upper floor joists will not remain at the initial mc of about 15 per cent but will try to swell to 19 or 20 per cent to the detriment of flooring and ceiling. When the joists finally settle down to a dried-out equilibrium, the disturbance, especially of the ceiling, is likely to remain with permanent cracks showing between wall and ceiling and in the plaster finish.

The doors and drawers in furniture would also tend to bind, but it would be wrong to attempt to make them fit properly during the drying out of the building since they would generally adjust themselves later.

The right balance of moisture in a new home will progressively improve but heating and ventilation should be carefully carried out. Even so, there are several situations in all types of building where it is extremely difficult to control the atmosphere to a satisfactory level and here it is essential that woodwork be constructed in such a way as to minimise unwanted movement.

Bowing of Doors

A problem that sometimes arises is the bowing of flush doors where different atmospheres on either side has caused one side to expand and the other side to contract. These include entrance doors, doors to airing cupboards in humid bathrooms, and interior doors giving access from a warm room to a cold one. When this occurs, usually the door has to be replaced, since the deformation is permanently set and any attempt to straighten the door, say by making a series of saw cuts across the expanded face and then applying an overlay on both sides after getting the door flat, is invariably uneconomic, and there is no assurance that the door will remain stable if the air conditions remain the same. There is also the situation where a home woodworker builds an extension to the property and fits a flush door which in service will be relatively exposed to the weather. In both situations it is essential to consider the reasons for bowing occurring and to take steps to remedy this when the door is hung.

Manufacturers of flush door attempt to provide a balanced construction in their products by ensuring that the materials on either side of a central core are identical in thickness, type, orientation, and moisture content, and such panels will remain flat provided the atmospheres on either side remain the same. However, this is not always possible, particularly in winter when air conditions can vary considerably on either side of an entrance door, and similar conditions could apply in say, a bathroom.

Since given air conditions equal an equilibrium moisture content for organic materials, obviously a difference of EMC on either side of a flush door will create a situation where one side of the door is attempting to shrink and the opposing side to swell. What is perhaps not so obvious is that if the EMC is the same on either side, but the temperatures are different, then there can be a temperature gradient through the thickness of the door, and this alone can cause thermal expansion on one side and contraction on the other.

Bowing due to thermal causes alone is not as a rule significant, but a temperature gradient through the thickness of a door usually results in a moisture gradient where moisture vapour diffuses from a high pressure level to a lower one. When a flush door is manufactured its moisture content is maintained at about 10 per cent, and if it has been properly looked after it will be at this level when installed. Since the air conditions in the average heated home is around this level as an equilibrium figure, any bowing that occurs in a door is often due to moisture transmission from the outside causing that side of the door to expand.

Flush doors vary in their core composition; some are provided with a skeleton infill, others are hollow, but all contain a lot of air between the outer surface skins. It can be seen that if the air on the outer side of a flush door is moist, the overall effect will be an expansion of the outer face of the door. This could happen in reverse, but only if the relative humidity on the inner side of the door was higher than that of the outside air.

It can be seen that if different relative humidities need to be held in check, then some form of vapour barrier must be applied to the door construction. The glue lines in plywood generally used to face external doors do to some extent offer a mild form of vapour barrier, but much depends upon the thickness of the glue line and often they do not entirely prevent vapour transmission. In any case, the outer plys can still absorb moisture. Where bowing of a flush door is known to be a possibility consideration should be given to the type of finish that is to be used on the door in order to make this an effective vapour barrier.

Normal paint and varnish systems are not very effective in the sense of being vapour resistant. Their primary function is to shed water and provide decoration and to a very great extent they allow the wood to breathe. Much depends, of course, on the amount of material applied and on its maintenance, but the problem in flush door bowing is not with liquid water but with vapour and this can progressively penetrate through thin surface films.

Where a varnish finish is proposed, the phenolic yacht varnishes are better than the alkyd types, but to be really vapour proof they must be applied in multiple coats and maintained annually. With paint finishes, a very effective vapour barrier is the use of aluminium sealer followed by the normal under coat and top coats. Whatever is applied must be the same for both sides of the door in order to maintain the balance. This is also important in the case of aluminium sealer because if this is applied only to one side of the door, vapour transmission from the opposing side can cause blistering of the final finish.

Efflorescence

It is perhaps pertinent to the question of moisture in buildings to mention efflorescence since from time to time this is mistaken for evidence of the presence of decay organisms in wood. Efflorescence derives from the Latin *efflorescere*; to bloom, and is a term used in chemistry to describe the changes which some crystals undergo when exposed to air. The salts contained in many materials are drawn to the surface and there they crystallize, the powdery surface bursting into 'flower'. Quite frequently, a householder will see these white flowers formed on the heads of nails used in the construction of wood roof joists, or occasionally on the wood itself, and conclude there is something sinister. Efflorescence is a harmless occurrence due to drying out of a material.

Condensation and Wood

When moist air is cooled below dew point, i.e. cooled to a temperature at which it cannot contain all the water originally present, and if the cooling is caused by contact with a colder surface, the surplus water appears as droplets on that surface as condensation.

In new houses it takes about six to twelve months for water from brick walls, concrete floors, and timber framing to evaporate, some of this moisture moves outwards through the walls and is lost to the outside air by natural ventilation, but much of it finds its way into the rooms. Apart from this aspect, air in a house can become moist because the occupants and some domestic appliances produce water vapour. Typical quantities of water vapour produced in the home are:

Adult (breathing)	0.1 litres per hour
Hot bath	1.5 litres per hour
Washing machine	3.0 litres per hour
Clothes drier	5.0 litres per hour
Hot shower	10.0 litres per hour

Water vapour is also generated in large quantities by gas stoves and paraffin heaters, so a lot of water may be 'stored' in the air each day as vapour. Steam, coming from service areas such as kitchens, laundries and bathrooms, travels to other rooms, due to differences in vapour pressures.

Where condensation is a problem and wood items and other materials are suffering in consquence, the answer is to remove the moisture laden air by ventilation and/or raising the temperature of any cold interior surface to a level above the room air dew point by heating. Control of temperature and humidity will make for better working conditions and will safeguard wood against excessive movement.

Tables

References

BARNES, D., ADMIRAL, L., PIKE. R.L., and MATHUR, V.N. Continuous system for the drying of lumber with microwave energy. *For. Prod. Journal* 26 (5); 31–42 1976.

BRITISH STANDARDS INSTITUTION: Nomenclature of commercial timbers including sources of supply; B.S. 881 & 589; London 1974.

BUILDING RESEARCH ESTABLISHMENT: PRINCES RISBOROUGH LABOROATORY: Timber Drying Manual. G.H. Pratt: Revised C.H.C. Turner. H.M.S.O. 1986.

BUILDING RESEARCH ESTABLISHMENT: PRINCES RISBOROUGH LABORATORY: A Handbook of Softwoods. H.M.S.O. 1977.

BUILDING RESEARCH ESTABLISHMENT: PRINCES RISBOROUGH LABORATORY: Handbook of hardwoods; revised by FARMER, R.H. HMSO London 1972.

BUILDING RESEARCH ESTABLISHMENT: PRINCES RISBOROUGH LABORATORY: The movement of timbers; Leaflet No 47; 1961. Kiln drying schedules; Technical Note No 37; 1969.

CHEN, P.Y.S., AND HELMER, W.A. Principles of dehumidification lumber drying. *For. Prod. Journal;* 32(5) 24–28; 1982.

CORKHILL, T. A Glossary of wood. Stobart & Son Ltd. London 1979.

EDLIN, H.L. What wood is that? Stobart & Son Ltd. London 1969.

HALL, G.S., HOOKS, R.A. and PLUMRIDGE, R.J. The Art of Timber Drying with Solar Kilns. Paper presented at Seminar on "Economic Criteria for the Selection of Woodworking Machinary and Plant Systems", Hannover, 19th May – 2nd June 1981.

HARRIS, R.A. & TARAS, M.A. Comparison of moisture content distribution, stress distribution, and shrinkage of red oak lumber dried by radio-frequency/vacuum drying process and a conventional kiln. *For. Prod. Journal;* 34(1) 44–54; 1984.

HOOKS, R.A.; Drying of British Wood: Timber Research & Development Association Research; Study TD/RS/2; 1977. High Wycombe.

Intermediate Technology Industrial Services (ITIS): An improved solar-kiln for drying timber; Project Bulletin; Rugby. 1984

JANE, F.W. The structure of wood: Adam & Charles Black; London 1956.
— The structure of wood. (2nd Edn., revised by Wilson, K. and White D.J.B.) London 1970. (see also Wilson/White, Anatomy of Wood).

KUBLER, H. AND HUNG CHEN, T. How to cut tree discs without formation of checks. *For. Prod. Journal* 24(7) 57–59; 1974.

LEE, A.W.C. AND HARRIS, R.A. Properties of red oak dried by radio-frequency/vacuum process and dehumidification process; *For. Prod. Journal* 34(5) 56–58; 1984.

LIESE, W. AND BAUCH, J. (Hamburg): On the closure of bordered pits in conifers; *Wood Science and Technology* Vol 1; 1–13; 1967.

LINCOLN, W.A. World Woods in Colour. Stobart & Son Ltd. London, 1986.

METEOROLOGICAL OFFICE: Averages and frequency distributions of humidity for Great Britain and Northern Ireland: *Climatological Memorandum* 103; 1961–70; 1976.

MITCHELL, H.L. How PEG helps the hobbyist who works with wood. U.S. Dept. of Agriculture; For. Prod. Lab. Madison; 1972.

MORÉN, R.E. Mo Och Domsjö AB, Örnsköldsvik, Sweden. Some practical applications of polyethylene glycol for the stabilisation and preservation of wood. Paper at British Wood Preserving Association Convention; Cambridge University; 1964.

OXLEY, T.A. When the moisture meter beats the oven; Trade leaflet; undated

PANSHIN, A.J.: DE ZEEUW, CARL: Text Book of Wood Technology: Volume 1: McGraw–Hill. 1964.

PLUMPTRE. R.A. Some thoughts on design and control of solar timber kilns. Paper to Wood Drying Workshop of IUFRO Division V Conference; Madison, U.S.A. 1983.

RIETZ, R.C. AND PAGE, R.H. Air drying of lumber; a guide to industry practices; U.S. Dept. of Agriculture, Forest Service Handbook No 402; 1971.

ST-LAURENT, A. Improving the surface quality of rip-sawn dry lumber. *For. Prod. Journal* 23(12) 1973.

TIMBER RESEARCH AND DEVELOPMENT ASSOCIATION: NOISE AND WOODWORKING MACHINERY; Timber Yard Operating Manual; Information Bulletin 3; 1971.

WENGERT, E.M. WEIK, B.R. SCHROEDER, J. AND BRISBIN, R. Practical drying techniques for yellow poplar flitches. *For. Prod. Journal* 347/8 39–44 1982.

WENGERT, E.M. Solar heated lumber dryer for small business; 1980. Virginia Cooperative Extension Service; Virginia Polytechnic Institute and State University.

WILSON, K. AND WHITE, D.J.B. The Anatomy of Wood: its diversity and variability. Stobart and Son Ltd, London 1986.

WRIGHT, G.W. Schedules and drying times covering several Australian timbers. C.S.I.R.O. Melbourne; Project S6; undated.

Index